# ENVISIONING THE

# 2020

## CENSUS

Panel on the Design of the 2010 Census
Program of Evaluations and Experiments

Lawrence D. Brown, Michael L. Cohen,
Daniel L. Cork, and Constance F. Citro, *Editors*

Committee on National Statistics

Division of Behavioral and Social Sciences and Education

## NATIONAL RESEARCH COUNCIL
*OF THE NATIONAL ACADEMIES*

THE NATIONAL ACADEMIES PRESS
Washington, DC
**www.nap.edu**

THE NATIONAL ACADEMIES PRESS     500 Fifth Street, NW     Washington, DC 20001

NOTICE: The project that is the subject of this report was approved by the Governing Board of the National Research Council, whose members are drawn from the councils of the National Academy of Sciences, the National Academy of Engineering, and the Institute of Medicine. The members of the committee responsible for the report were chosen for their special competences and with regard for appropriate balance.

The project that is the subject of this report was supported by contract no. YA1323-06-CN-0031 between the U.S. Census Bureau and the National Academy of Sciences. Support of the work of the Committee on National Statistics is provided by a consortium of federal agencies through a grant from the National Science Foundation (No. SES-0453930). Any opinions, findings, conclusions, or recommendations expressed in this publication are those of the author(s) and do not necessarily reflect the views of the organizations or agencies that provided support for the project.

International Standard Book Number 13: 978-0-309-15115-3
International Standard Book Number 10: 0-309-15115-5

Additional copies of this report are available from the National Academies Press, 500 Fifth Street, NW, Lockbox 285, Washington, DC 20055; (800) 624-6242 or (202) 334-3313 (in the Washington metropolitan area); Internet, http://www.nap.edu.

Printed in the United States of America

Suggested citation: National Research Council. (2010). *Envisioning the 2020 Census*. Panel on the Design of the 2010 Census Program of Evaluations and Experiments. Lawrence D. Brown, Michael L. Cohen, Daniel L. Cork, and Constance F. Citro, eds. Committee on National Statistics, Division of Behavioral and Social Sciences and Education. Washington, DC: The National Academies Press.

# THE NATIONAL ACADEMIES
*Advisers to the Nation on Science, Engineering, and Medicine*

The **National Academy of Sciences** is a private, nonprofit, self-perpetuating society of distinguished scholars engaged in scientific and engineering research, dedicated to the furtherance of science and technology and to their use for the general welfare. Upon the authority of the charter granted to it by the Congress in 1863, the Academy has a mandate that requires it to advise the federal government on scientific and technical matters. Dr. Ralph J. Cicerone is president of the National Academy of Sciences.

The **National Academy of Engineering** was established in 1964, under the charter of the National Academy of Sciences, as a parallel organization of outstanding engineers. It is autonomous in its administration and in the selection of its members, sharing with the National Academy of Sciences the responsibility for advising the federal government. The National Academy of Engineering also sponsors engineering programs aimed at meeting national needs, encourages education and research, and recognizes the superior achievements of engineers. Dr. Charles M. Vest is president of the National Academy of Engineering.

The **Institute of Medicine** was established in 1970 by the National Academy of Sciences to secure the services of eminent members of appropriate professions in the examination of policy matters pertaining to the health of the public. The Institute acts under the responsibility given to the National Academy of Sciences by its congressional charter to be an adviser to the federal government and, upon its own initiative, to identify issues of medical care, research, and education. Dr. Harvey V. Fineberg is president of the Institute of Medicine.

The **National Research Council** was organized by the National Academy of Sciences in 1916 to associate the broad community of science and technology with the Academy's purposes of furthering knowledge and advising the federal government. Functioning in accordance with general policies determined by the Academy, the Council has become the principal operating agency of both the National Academy of Sciences and the National Academy of Engineering in providing services to the government, the public, and the scientific and engineering communities. The Council is administered jointly by both Academies and the Institute of Medicine. Dr. Ralph J. Cicerone and Dr. Charles M. Vest are chair and vice chair, respectively, of the National Research Council.

**www.national-academies.org**

# Acknowledgments

The Panel on the Design of the 2010 Census Program of Evaluations and Experiments (CPEX) wishes to thank the many people who have contributed to our work. The panel is indebted to former Deputy Director Preston Jay Waite's support during the developmental stage of the study and to the continued support from Daniel Weinberg, assistant director for decennial census and American Community Survey. Deborah Bolton and Randall Neugebauer ably served as principal liaisons between the Census Bureau and the panel, assisted by Gary Chappell, Joyce Price, Jennifer Reichert, and Donna Souders. All of these Census Bureau staff were responsive to requests from the panel for help in the development of agendas, the collection of supporting materials, and assistance with other logistical details. They have been a pleasure to work with.

Also, a number of Census Bureau personnel provided extremely useful presentations and supporting materials during the panel's meetings in April 2007, July 2007, April 2008, July 2008, November 2008, and February 2009. In this regard, we would like to thank Teresa Angueira, Michael Bentley, Larry Cahoon, Joan Hill, Elizabeth Martin, Mary Mulry, Manuel de la Puente, Jennifer Reichert, Courtney Reiser, Annetta Clark Smith, and Frank Vitrano. Many of these individuals, along with others, also contributed to three very productive small group meetings of the panel (held in July 2007), and here we would like to mention interactions with Nancy Bates, William Bell, Sharon Boyer, Mary Destasio, Donna Kostanich, Laurel Schwede, Jennifer Tancreto, and Jim Treat. We would like to single out Jim Treat to thank him for his help in learning about a number of disparate areas of census planning, experimentation, and evaluation.

Our panel also benefited from the contributions of a number of outside speakers at our plenary meetings who gave generously of their talents. We thank Mick Couper, University of Michigan, and John Czajka, Mathematica Policy Research, Inc., for their thoughtful presentations on Internet data collection and administrative records, respectively. Rosemary Bender of Statistics Canada provided useful insight on census research in Canada, contributing to a discussion organized by panel member Ivan Fellegi. Finally, Marios Hadjieleftheriou of AT&T spoke about the possible structure of a master trace system and the use of such systems in industry.

We would like to thank Constance Citro, along with Jay Waite, for developing the study. In addition, Connie participated actively in our panel meetings and helped with writing and editing this report, providing her usual insightful comments and suggestions. Her knowledge and experience of census history and procedures proved to be an essential and invaluable addition to our deliberations. We would also like to thank Meyer Zitter, consultant to the panel, for also providing important, experienced perspectives on the census and in particular for his advice on the potential use of administrative records in census taking. We express our gratitude to Agnes Gaskin, who has seen to it that the meetings of the panel run as smoothly as possible, facilitating the travel and other needs of the panel members, and formatting this report. The panel is also indebted to Christine McShane, who provided expert technical editing of the draft report.

Michael Cohen and Daniel Cork served as extraordinarily effective and essential co-directors for the work of the panel. They admirably fulfilled their responsibility for compiling, organizing and presenting most of the background material and basic research appearing in the report, and for skillfully organizing and drafting the report so as to integrate the panel's opinions and concerns along with our individual contributions. They were responsible from the side of the panel in arranging our cooperative interactions with the Census Bureau, which have been essential in our deliberations. And their experience and common sense helped keep us on track and focused as a panel during meetings and during the process of writing our interim and final reports.

Finally, it has been a pleasure interacting with a very talented, energetic, and collaborative panel as we considered plans for evaluation of the 2010 census and for research and development to be carried out in the next decade to achieve as cost effective as possible a census in 2020.

This final report has been reviewed in draft form by individuals chosen for their diverse perspectives and technical expertise, in accordance with procedures approved by the Report Review Committee of the National Research Council. The purpose of this independent review is to provide candid and critical comments that will assist the institution in making the published report as sound as possible and to ensure that the report meets institutional standards for objectivity, evidence, and responsiveness to the study charge. The review comments and draft manuscript remain confidential to protect the integrity of the deliberative process. We thank the following individuals for their participation in the review of this report: Betsy Ancker-Johnson, vice president (retired), General Motors Corporation; John L. Czajka, Mathematica Policy Research, Inc., Washington, DC; Shoreh Elhami, GIS director, Delaware County Auditor's Office, Delaware, OH; Benjamin King, statistical consultant, Durham, NC; Daniel B. Levine, Statistics, Westat, Inc., Rockville, MD; Betsy Martin, consultant, Alexandria, VA; C. Matthew Snipp, Department of Sociology, Stanford University; and Alan M. Zaslavsky, Department of Health Care Policy, Harvard Medical School.

In addition, we thank the following individuals who served as reviewers for *Experimentation and Evaluation Plans for the 2010 Census: Interim Report*, which is reprinted in Part II of this volume: C.A. (Al) Irvine, consultant, San Diego, CA; Benjamin King, statistical consultant, Durham, NC; J. Michael Oakes, Division of Epidemiology, University of Minnesota, Minneapolis; Joseph Salvo, Population Division, New York City Department of City Planning, New York City, NY; Robert Scardamalia, Center for Research and Information Analysis, Department of Economic Development, Albany, NY; Frederick J. Scheuren, consultant, Alexandria, VA; and Judith M. Tanur, Department of Sociology and Behavior, State University of New York, Stony Brook. Finally, we thank the reviewers of the panel's February 2009 letter report, which is reprinted as Part III of this volume: Barbara A. Bailar, independent consultant, Washington, DC; John L. Czajka, Mathematica Policy Research, Inc., Washington, DC; C.A. (Al) Irvine, consultant, San Diego, CA; Benjamin King, statistical consultant, Durham, NC; and Colm A. O'Muircheartaigh, Harris Graduate School of Public Policy Studies, The University of Chicago.

Although the reviewers listed above provided many constructive comments and suggestions, they were not asked to endorse the conclusions or recommendations nor did they see the final draft of the report before its release. The review of this report was overseen by Samuel H. Preston, Population Studies Center, University of Pennsylvania, and William F. Eddy, Department of Statistics, Carnegie Mellon University. Appointed by the National Research Council, they were responsible for making certain that an independent examination of this report was carried out in accordance with institutional procedures and that all review comments were carefully considered. Responsibility for the final content of this report rests entirely with the authoring committee and the institution.

Lawrence D. Brown, *Chair*
Panel on the Design of the 2010 Census
Program of Evaluations and Experiments

# Contents

# List of Figures

*xvii*

# List of Tables

# List of Boxes

# Part I

# Final Report

# Summary

THE DECENNIAL CENSUS IS A CORNERSTONE of the U.S. federal government and its statistical system. The census's constitutionally mandated purpose is to provide population counts to reapportion the U.S. House of Representatives every 10 years; a closely related purpose is to provide block-level population counts by age, race, and ethnicity to draw the boundaries of congressional and legislative districts. In addition, census information is used to allocate hundreds of billions of dollars of federal funds to states and localities; as the basis for postcensal population estimates for states, counties, and cities; to support federal, state, and local government program planning; to provide sampling frames and population weights for household surveys; and to provide denominators for such vital statistics as birth and death rates by age, gender, race, and ethnicity.

To address these information needs, the Census Bureau is preparing to conduct the nation's 23rd decennial census. In 2010, the Census Bureau will collect basic information for each U.S. resident, including name, gender, age, date of birth, race, ethnicity, relationship to principal respondent, number of people in the housing unit, and whether the housing unit is owned or rented. The additional socioeconomic information that was previously collected on the census long-form questionnaire for a sample of households is now collected by the continuous American Community Survey (ACS), resulting in a short-form-only 2010 census.

Census planning is itself a major effort, which since the 1950 census has involved extensive research and development over the better part of each decade. The planning culminates in a dress rehearsal for the immediately upcoming census, followed by experiments and evaluations conducted as part of that census to begin the planning for the next census.

The Census Bureau recognized that the experiments and evaluations conducted for the 2000 census did not provide an adequate basis for 2010 census planning. The Bureau therefore requested a study from a panel convened by the National Academies Committee on National Statistics to review the 2010 Census Program of Experiments and Evaluations (CPEX) so that it would provide a strong basis for 2020 census planning. The panel was charged to:

> consider priorities for evaluation and experimentation in the 2010 census. [The panel] will also consider the design and documentation of the Master Address File and operational databases to facilitate research and evaluation, the design of experiments to embed in the 2010 census, the design of evaluations of the 2010 census processes, and what can be learned from the pre-2010 testing that was conducted in 2003–2006 to enhance the testing to be conducted in 2012–2016 to support census planning for 2020. Topic areas for research, evaluation, and testing that would come within the panel's scope include questionnaire design, address updating, nonresponse follow-up, coverage follow-up, unduplication of housing units and residents, editing and imputation procedures, and other census operations.

Together, the panel's interim and letter reports (included in this volume as Parts II and III) fulfill the core mandate of this charge to assess the 2010 CPEX program. In those reports, the panel generally concluded that the specific experiments chosen by the Census Bureau for the 2010 CPEX are not likely to inform changes that will substantially affect either census cost or quality in 2020, and so largely squander a valuable testing opportunity. Moreover, the panel concluded that the ability to evaluate the 2010 census (apart from or in addition to the formal CPEX evaluations) will depend on the Census Bureau's ability to retain input and output data from each 2010 census operation, for later linkage and analysis.

## RESEARCH AND DEVELOPMENT FOR A
## COST-EFFECTIVE 2020 CENSUS

This final report builds on the panel's earlier work by discussing census experimentation and evaluation in a broader context. In particular, this report describes a strategy for research and development (R&D) leading to a 2020 census design that controls costs, maintains quality, and adapts to major social and technological changes. An effective research and development strategy for the 2020 census should keep a goal of containing costs while maintaining quality; it should also recognize that new technical advances may not only reduce costs, but also drive them up, so that research and effective cost modeling are necessary to determine how best to use such advances for the most cost-effective census.

## Census Costs

Since 1970, when the current approach of a mailout-mailback enumeration, based on a master address list, with in-person follow-up of addresses for which a questionnaire is not returned, was first used, the real dollar per housing unit cost (2009 dollars) of the U.S. census has increased by more than 600 percent—it was $17 in 1970 (about the same as in 1960), $33 in 1980 (91 percent increase over 1970), $43 in 1990 (30 percent increase over 1980), and $70 in 2000 (63 percent increase over 1990). It is anticipated to be as much as $115 in 2010 (64 percent increase over 2000). In contrast, the real dollar per housing unit cost of recent Canadian censuses (in 2008 Canadian dollars) has remained constant at $39–$40 for the 1996, 2001, 2006, and 2011 censuses.

The unprecedentedly high per housing unit cost expected for 2010 has resulted in part from planned long-term investments—improvements to the Census Bureau's TIGER geographic database, which is essential for accurate geographic coding and mapping of addresses, and the testing and implementation of the continuous ACS. High costs have also resulted from a failed initiative to automate the nonresponse follow-up operation, which will use traditional paper-and-pencil questionnaires instead of handheld computers as originally planned. Moreover, contributing to cost increases over the entire 40-year span has been the accretion of operations to address special situations, which have not been fully evaluated to determine their benefits and costs. Underlying this accretion and the failure to achieve major steps forward on a timely basis, such as the nonresponse follow-up automation initiative, is a largely incremental approach to census design and a research and development program that is often unfocused and ineffective.

## Census Quality

The overall quality of the census enumeration using the current mailout/mailback methodology has improved between 1970 and 2000, to the point at which major continued improvement is unlikely. The net undercount of the population (measured by the demographic analysis method) decreased from 2.7 percent in 1970 to 0.1 percent in 2000. Moreover, the difference in net undercount rates between African Americans and all others decreased from 4.3 percentage points in 1970 to 3.1 percentage points in 2000. These figures mask large numbers of offsetting errors of omission of people who should have been counted and duplicate or other types of erroneous enumeration of people who should not have been counted.

Some programs put in place to address coverage errors, such as a recheck of housing units initially classified as vacant, have contributed to reducing net undercount (they have also contributed to erroneous enumerations).

Other programs, sometimes put in place with no advance testing, such as a parolee and probationer check in 1990, have contributed little to coverage improvement. Given trends that make census-taking more difficult, it is unlikely that the increased costs anticipated for the 2010 census will achieve much, if any, further increase in quality. In order to lead to successful quality improvements for 2020 any additional expenditures would, at a minimum, need to be properly focused and carefully researched and planned.

### Social and Technological Change

Changes in the U.S. population since 1970 that have made census-taking more difficult are well known. They include increased immigration, including illegal immigration, and therefore the existence of communities that are wary of cooperating with the census and in which English is not the primary language. Changing norms in residence and living arrangements have also complicated the concept of a single, usual place of residence for many people, including children in joint custody, people with seasonal homes, commuter workers or couples who maintain a separate workweek residence, and others. More broadly, the public's willingness to respond to surveys has declined significantly over the past three decades. Given these factors, an incremental approach to census design and an unfocused R&D program are likely to result in a continuation of the pattern of escalating census costs in 2020 with little or no improvement in quality.

Nevertheless, important dynamics that are dramatically changing the environment in which the census is taken offer both challenges and opportunities to make the 2020 census markedly more cost-effective. Administrative record files from government assistance and other programs are improving in quality and completeness, the Internet is now a ubiquitous means for communication, and the proportion of household business that is carried out through the ordinary mail is shrinking. It is difficult to paint a reliable picture now of the United States in 2020, but it is easy to conceive of the Internet as being the primary method for communication and conducting household business for the great majority of residents, and that a large fraction of U.S. households may opt not to receive ordinary mail or may ignore much or all of it. It is also likely that administrative records will provide timely, high-quality, and inexpensive information that would be a useful input to a variety of census operations. Clearly, the design of the 2020 census should be dramatically different to accommodate such changes.

## TOWARD A NEW VISION FOR THE 2020 CENSUS

To escape the pattern of incremental and often unfocused research and development for the next census leading to escalating costs with diminishing returns on quality, the Census Bureau will need to completely overhaul its approach for planning the 2020 census. Planning for 2020 must begin with a set of clear goals for census costs and quality, a limited set of strategic visions that are likely to meet those goals, and an R&D program that is focused on determining which vision has the most promise and how best to implement it. The goal for census costs should recognize that an increase in real dollar per housing unit cost for 2020 over 2010 would be unjustified in comparison with the experience of other developed nations and unacceptable in a time of fiscal imbalance.

> *Recommendation 2.1:* **The Census Bureau should approach the design of the 2020 census with the clear and publicly announced goal of reducing the inflation-adjusted per housing unit cost to that of the 2000 census (subtracting the cost of the 2000 census long-form sample), while holding coverage errors (appropriately defined) to approximately 2000 levels.**

To be clear, we do not make this recommendation to suggest cutting census costs simply for the sake of cost-cutting. The objective is a more efficient census that, for example, uses lower cost options like the Internet and administrative records to supplement or replace expensive follow-up operations without compromising census quality; effective research and improved cost modeling (see below) are critical to determining to what extent these efficiencies can be realized. The reason for setting a cost goal that is stark, ambitious, and public is to convey commitment to a focused 2020 census planning process.

This 2020 census cost goal should exclude the costs of the ACS, which will increasingly take on a life of its own as a source of intercensal data on a wide range of population characteristics.

> *Recommendation 3.1:* **The Census Bureau should immediately develop a limited number of strategic visions for the 2020 census that are likely to meet its announced goals for costs and quality. By strategic visions, we mean start-to-finish strategies for conducting all major census operations in order to confront looming threats and implement new technologies. In addition, evaluation of each vision should include thorough review of the costs and benefits of individual census operations relative to the announced cost and quality goals, to determine whether operations that are not demonstrably cost-effective may be eliminated or scaled back or whether new technologies or practices might usefully be introduced.**

> *Recommendation 3.2:* **The Census Bureau should develop its 2020 census research and development program by identifying a handful of research questions whose resolution will determine**

which of the visions of the next census are feasible and cost-effective and which are not. Priorities for evaluations of the 2010 census should be arrived at consistent with this research program, as should priorities for experiments and tests in the 2011–2018 period.

*Recommendation 4.7:* The Census Bureau's planning for the 2020 census, particularly for research in the period 2010–2015, should be designed to permit proper evaluation of significant innovations and alternatives to the current decennial census design that will accomplish substantial cost savings in 2020 without impairing census quality. Otherwise, the census design in 2020 will either be an incremental change from that in 2010 with increased costs, or the Census Bureau may be compelled to implement a poorly evaluated and tested alternative design under severe time and cost constraints with a risk of substantially reduced quality. All involved, including Congress and the administration, should recognize that substantial cost savings in 2020 can be achieved only through effective planning over the course of the 2010–2020 decade and should fund and pursue research efforts commensurately.

## PROVIDING TOOLS FOR CENSUS R&D

To make possible a truly focused R&D program for 2020 census planning, the Census Bureau needs to take immediate steps to develop the necessary tools for effective planning and evaluation. These tools include an improved, transparent cost model of census operations and well-documented data from all 2010 census operations in formats suitable for research and evaluation.

*Recommendation 3.3:* In order to provide early indications of the costs of competing visions for the 2020 census and to support effective planning throughout the decade, the Census Bureau should develop and validate a detailed cost model that not only represents the 2010 census, but also accommodates novel approaches to census-taking, including the use of data capture via the Internet and automated telephone systems, the use of handheld devices in nonresponse follow-up, the use of administrative record information for some types of nonresponse follow-up cases, and innovative mechanisms for reducing the costs of updating the Master Address File between 2010 and 2020. This cost model should be able to assess the implications of introducing specific changes to an existing design, singly and in combination, and to distinguish between the direct and indirect cost effects of specific changes. This cost model should be thoroughly documented and transparent so that the Census Bureau can obtain the benefit of expert advice on cost-effective improvements to census operations.

*Recommendation 3.4:* The Census Bureau should retain sufficient input and output data, properly linked and documented, from each 2010 census operation to permit adequate evaluation of the contribution of the operation to census costs and data quality to feed into 2020 census planning. For this purpose, the Census Bureau should either establish an internal group or hire a contractor with database management expertise. This group would have the responsibility of retaining and documenting sufficient data from the 2010 census to be able to comprehensively represent the functioning of all census operations. Such a group would also have the responsibility of assisting Bureau research staff, using current database management tools, to produce research files to support the assessment of analytic questions concerning aspects of the 2010 census.

## REVITALIZING THE RESEARCH INFRASTRUCTURE

To support a focused, goal-oriented R&D program for the 2020 census and, more broadly, to improve its research capabilities, the Census Bureau needs to take immediate steps to revitalize its research infrastructure. These steps include thorough assessment of the Census Bureau's posture toward R&D compared with other data collection organizations and the development of means for research to break out of existing organizational "silos."

*Recommendation 4.1:* The Census Bureau should comprehensively review the research and development practices and organization in other national statistics offices and in survey organizations in academia and the private sector, with the goal of modernizing and strengthening the Bureau's own research and development program. Such a review should include assessments of and recommendations about:

- How to organize and direct basic and applied methods research to best serve the decennial census and other Census Bureau programs;

- How to organize information technology and database management to best serve research and operations, including how to manage the development of new technologies and ensure access to adequate expertise in these technical areas;

- How to operate collaborative project teams to facilitate timely innovation;
- How to ensure adequate training in survey methods and related fields;
- How to achieve extensive and intensive interaction with external research organizations and academic departments so that Census Bureau researchers and methodologists can benefit from related research work and ideas elsewhere; and
- How to fund and establish priorities for research and applied methodology work.

*Recommendation 4.4:* The Census Bureau should put in place incentives and structures so that research is fully integrated and collaborative not only across programs, but also with operational planning. Research should be responsive to operational needs, and, in turn, research findings should play a primary role in informing operational decision making.

Necessary steps also include reestablishing the position of Associate Director of Methodology and Standards, reestablishing a strong Center for Survey Methods under that associate director, integrating research across programs and with operational planning, integrating census and ACS research, and renewing and refreshing mechanisms for obtaining outside expert advice.

*Recommendation 4.2:* To carry out the findings from the review recommended above, the Census Bureau should consider reestablishing and filling an associate director–level executive staff position to head the statistical and survey research activities at the Census Bureau, with authority to organize the Bureau's research and applied methods activities.

*Recommendation 4.3:* The Census Bureau should give greater emphasis to survey methodology. One possibility for doing so would be to establish a core survey methods research center, staffed by full-time survey researchers and headed by a nationally recognized expert in census and survey data collection instruments. Such a high-profile center could give priority to research on making effective initial contacts with census and survey respondents, including those made with new technologies.

*Recommendation 4.5:* The Census Bureau should integrate decennial census and American Community Survey research—for example, by using the ACS methods panel as a test bed for the Internet and other data collection methods to consider in the census and by matching census and ACS records to evaluate coverage in both programs. To support comparative census-ACS research and to inform users, the Census Bureau should carry out analyses that explore, at both the aggregate level and the level of individual households, the degree of differences and the source of differences in demographic characteristics and residence between the ACS and the decennial census.

*Recommendation 4.6:* The Census Bureau should renew and augment mechanisms for obtaining external expertise from leading researchers and practitioners in survey and census methodology and in relevant computer science fields. These mechanisms might include (1) a more active census professional advisory committee program in which the members have an opportunity to work more closely with Census Bureau staff in developing and evaluating ideas for improved census and survey methods; (2) increased opportunities for sabbaticals at the Bureau for university faculty and other short-term appointments for both senior- and junior-level (graduate student) academics at the Census Bureau; (3) increased opportunities for sabbaticals for Census Bureau staff at academic institutions and private-sector survey organizations; (4) the awarding of design contracts early in the decade to support research and development of innovative technologies for census and survey data collection and processing; and (5) more effective use of contracting processes to obtain expert services.

## 2020 R&D FOR STAGES OF CENSUS-TAKING

The strategic visions developed for the 2020 census will include different combinations of approaches for more cost-effective address list development, initial contact with and response mechanisms for household and group quarters residents, and follow-up operations. The panel suggests targets of opportunity for focused R&D for each of these stages of census-taking.

### Address List Development

A priority for determining a more cost-effective strategy for developing the Master Address File (MAF) between 2010 and 2020 is to conduct a comprehensive evaluation of the accuracy and costs of the 2010 Decennial Master Address File (DMAF) at each stage of its development. This evaluation should determine, by comparison with results from the 2010 census, the 2010 ACS, and the 2010 Census Coverage Measurement, which operations—Postal Service updates, Local Update of Census Addresses, block canvass, Vacant/Delete Check, and others—resulted in accurate addresses (occupied and vacant), which operations added nonresidential and nonexistent units, and which operations erroneously deleted valid addresses. The evaluation should examine the comparative performance of various address operations for geographic areas, types of neighborhoods, and types of structures, such as single-family residences, small multiunit structures, and large multiunit structures. It should simulate the effects on census costs and accuracy of deleting one or more address operations nationwide and for types of areas and structures.

Research and development should investigate the use of administrative records from government programs to replace more expensive address updating operations in some areas and the use of information on population and housing trends from the ACS and the Census Bureau's population estimates program to target areas in which address canvassing may not be needed. R&D should also investigate the costs and benefits of continuous updating of the MAF with the cooperation of some technologically savvy state and local governments and pilot-test a continuous updating program in selected areas. A continuously updated MAF would have benefits throughout the decade for the ACS and other household surveys.

### Initial Contact and Response

There are a number of ways of contacting households to let them know how and when to respond to the census, including not only the traditional mailing of a questionnaire but also mailing a postcard, sending an e-mail, or leaving a voice message. Similarly, Internet and interactive voice response could become much more prevalent modes of household response. There are issues of confidentiality protection and ensuring the ability to link a response to a particular address, yet other countries have solved these problems and are making extensive use of Internet response to their censuses.

Only about 3 percent of people reside in institutions and other kinds of group quarters, but in some local areas they constitute a large proportion of the population. Group quarters enumeration poses special challenges—first, of identifying all group quarters and then of gaining access and accurately enumerating group quarters residents. An early priority for R&D on more effective group quarters enumeration should be to investigate the use of administrative records to enumerate the residents of institutional group quarters, such as nursing homes, prisons, and college dormitories.

## Follow-Up

Nonresponse follow-up and other field operations, such as the Vacant/ Delete Check, are time-consuming and add substantially to the costs of the census. Focused R&D for a cost-effective 2020 census should address ways to minimize field operations by making more extensive use of administrative records, as well as ways to reduce the costs of fieldwork, assuming that some level of in-person follow-up will continue to be required.

R&D on minimizing field operations should analyze 2000 and 2010 information to determine the likely effects on costs and data quality of using administrative records to replace some or all of nonresponse follow-up in some or all geographic areas. Options to consider include the use of administrative records to replace current last-resort procedures in late stages of follow-up; the use of administrative records after one or two attempts at in-person nonresponse follow-up; and the use of administrative records at the outset so that nonresponse follow-up is conducted only when there is no match for an address to a record. Options for reducing the cost of in-person follow-up include reducing the number of contact attempts before using last-resort procedures (or administrative records) and the use of handheld technology in place of paper questionnaires.

## Cost-Benefit Review of Individual Census Procedures

A close examination of the modern U.S. census reveals the accretion of many individual procedures to address special situations for address list development, initial contact, nonresponse follow-up, and telephone and in-person coverage improvement follow-up. To satisfy a goal of a substantial reduction in real dollar per housing unit cost for the 2020 census, the Census Bureau will need not only to test and develop major changes in census operations, but also to examine thoroughly each individual procedure to determine its contribution to census costs and data quality.

This exercise will require the detailed, transparent cost modeling recommended above, so that stakeholders can be informed regarding trade-offs between cost savings and possible adverse effects on completeness of census coverage. While significant deterioration of data quality should not be contemplated, elimination of procedures that only marginally benefit quality and have significant costs should not be ruled out a priori. The U.S. census needs to be of high quality; it also needs not to consume disproportionate resources. Focused R&D that builds from strategically chosen evaluations of 2010 census results and is supported by detailed cost modeling is the necessary ingredient to achieving a new vision for a cost-effective 2020 census.

# – 1 –

# Introduction

THE U.S. DECENNIAL CENSUS IS INTEGRAL to the regular realignment of power and resources mandated by Article I, Section 2, of the U.S. Constitution. In addition to apportioning the U.S. House of Representatives, the census provides the basic population count information used to redraw congressional and other legislative districts. Census data also serve a wide variety of other purposes, including the allocation of government funds, the construction of frames and weights for many surveys, and the development of numerous statistical measures and indexes.

The 2010 census will be the nation's 23rd decennial enumeration. Historically, the decennial census has been characterized by both change and continuity. Marshals on horseback have given way to temporary enumerators working for a permanent Census Bureau, methods based solely on personal visits to those relying principally on the mail, and a long-form sample of respondents providing additional socioeconomic data to the continuous American Community Survey. The machinery of census-taking has changed as well, from handwritten ledgers to Hollerith punch cards to the first UNIVAC computer to modern document-scanning technology. Yet the 2010 census—like all of its predecessors—will inherit some key concepts that have been part of the census experience since the first enumeration in 1790. The de jure standard of trying to associate people with their single place of "usual residence" that will be used in the 2010 census is directly descended from the "usual place of abode" standard dictated by the first U.S. Congress for the 1790 census (1 Stat. 101). The act that authorized the first census in 1790 also made response to census inquiries mandatory of all residents, as is still the case in modern statute.

Systematic research and development (R&D) to improve census methodology, pretest changes in procedures, and evaluate census results is an integral part of the modern census. Such R&D was not possible in the early decades of census-taking because there was no permanent census office with appropriate staffing, and statistical methods for testing and evaluation were in their infancy. An exception that illustrates the importance of research and testing occurred for the 1850 census: a temporary Census Board, created in 1849, solicited outside expert advice that resulted in marked improvements in the design of the census forms that were used by U.S. marshals to collect the data. The improvements were based on an 1845 census of Boston directed by Lemuel Shattuck, a founder of the American Statistical Association (Hacker, 2000a).

A permanent Census Bureau was established in 1902, and the 1910 census included perhaps the first major *experiment* conducted in conjunction with a U.S. census. In order to boost awareness of the 1910 census and save interviewing time, census enumerators in large cities (100,000 population or greater) distributed "advance schedules" a few days prior to the beginning of the actual count. Although respondents were encouraged to fill out the schedule in advance of the enumerator visit, enumerators were not permitted to simply accept a filled-out form as a response to the census; simply "prepar[ing] the way for the enumerator by announcing his approaching visit and informing the people precisely of the questions to be answered" was the primary aim of the experiment, while potential time savings and gains in accuracy were described as the "secondary object" (U.S. Department of Commerce and Labor, 1911:115–116). Three decades later, the Census Bureau's plan for the 1940 census included the first formal, structured set of *evaluations* of census content and quality in a decennial census; the procedural history of the 1940 census commissioned by the Bureau describes a series of analyses of the data quality for individual question items as well as limited work to estimate the level of underenumeration in the census (Jenkins, 1983:96–104). The 1950 census included the first systematic set of pretests and experiments to plan the enumeration, along with the first set of experiments conducted as part of the census and a large number of post-census evaluations (Goldfield and Pemberton, 2000a).

The Census Bureau uses several terms to delineate its census R&D activities, including pretest or test, experiment, and evaluation. Although usage is not always consistent, in this terminology, "tests" involve collection of data from respondents between censuses. Some tests shake down operations and uncover potential problems in the field; other tests are more properly termed experiments, in that they involve designed comparisons of alternative questionnaires, field procedures, and other aspects of the census. The Bureau, however, typically reserves the term "experiment" for instances when data collection to compare alternative methods is conducted as part and parcel of the census itself. Finally, "evaluations" involve after-the-fact comparisons of census results with other surveys or data sources. Many evaluations assess operational or data quality by using information that is tracked and collected as census operations take place. Other evaluations, notably the postenumeration survey that is commonly used to assess coverage errors in the census, involve original data collection. Our panel suggests that fixation on these labels is not as helpful as thinking in terms of census "research and development" generally.

Since the 1950 census, some program of both experimentation and evaluation has been a fixture of the decennial census process. These formal programs have developed into major enterprises: not counting studies directly related to the postenumeration survey and coverage measurement, the 2000 census included a slate of 91 formal evaluations (pared from an original list of 149 possibilities), 5 formal experiments, and 3 studies based on qualitative, ethnographic observation of the census. No exception from its recent predecessors, the upcoming 2010 census has been designed to include a formal slate of experimentation and evaluation; early in the planning process, the Census Bureau dubbed this effort the 2010 Census Program of Evaluations and Experiments, or CPEX for short.

## 1–A   THE PANEL, ITS CHARGE, AND PREVIOUS REPORTS

The program evaluations of the 2000 census were sharply criticized by the National Research Council (2004a:Finding 1.11) Panel to Review the 2000 Census as "slow to appear, [often] of limited value to users for understanding differences in data quality[, and] of limited use for 2010 planning." Recognizing that the experiments and evaluations of the 2000 census did not provide an adequate basis for 2010 census planning, the U.S. Census Bureau requested that the Committee on National Statistics (CNSTAT) of the National Research Council convene this Panel on the Design of the 2010 Census Program of Evaluations and Experiments in early 2007. The panel was charged to:

> review the U.S. Census Bureau's program of research, evaluation, and experimentation for the 2010 census. The panel will consider priorities for evaluation and for evaluation and experimentation in the 2010 census. It will also consider the design and documentation of the Master Address File and operational databases to facilitate research and evaluation, the design of experiments to embed in the 2010 census, the design of evaluations of the dress rehearsal and 2010 census processes, and what can be learned from the pre-2010 testing that was conducted in 2003–2006 to enhance the testing to be conducted in 2012–2016 to support census planning for 2020. Topic areas for research, evaluation, and testing that would come within the panel's scope include

questionnaire design, address updating, nonresponse follow-up, coverage follow-up, unduplication of housing units and residents, editing and imputation procedures, and other census operations.

This is the panel's third and final report. At the Census Bureau's request, the panel issued a first, interim report in late 2007 (National Research Council, 2008b) with the express intent of reviewing an initial list of general topics prepared by the Census Bureau's CPEX staff to identify key priorities. In that report, we focused in particular on topics for experimentation that were likely to improve census quality, reduce census costs, and provide a basis for further research during the 2010–2020 decade. Accordingly, we identified two areas that were either entirely absent from the Bureau's initial list or mentioned only in limited fashion: the use of Internet data collection and the use of administrative records for a variety of census purposes. We further recommended that an alternative questionnaire experiment—a staple of previous census experimentation programs—center primarily on the presentation of residence concepts and accurate elicitation of household counts.

Subsequently, Census Bureau staff briefed the panel about its chosen roster of experiments and evaluation studies for 2010; we describe the Bureau's roster in more detail in Appendix B. In brief, in its selections, the Census Bureau chose not to pursue any form of Internet data collection experiment and also declined to study uses of administrative records other than their limited use in the 2010 Coverage Follow-Up operation. As expected, an alternative questionnaire experiment was chosen for the 2010 experiment roster, but its focus is almost exclusively on subtle and hard-to-distinguish differences in the wording and presentation of race and Hispanic-origin questions. The panel had remaining concerns about the Bureau's selected research priorities. Moreover, great uncertainty surrounded the progress of the Census Bureau's "replanning" effort following the discovery of major problems with its Field Data Collection Automation contract; it was unclear to the panel (and other external observers, such as the U.S. Government Accountability Office, 2008, 2009b,d) whether the operational control systems that manage the flow of information during the census process would be adequately tested prior to deployment. The panel was also concerned that, in the haste to simply get operational systems in a functional state, the Bureau might not give proper attention to archiving data or creating "spigots" of operational data in order to enable evaluation of census processes. Because of these concerns, the panel issued a second report in February 2009 (National Research Council, 2009a) in the form of a letter to then-acting Census Bureau Director Thomas Mesenbourg. The letter report provides the panel's detailed critique of the four experiments known to the panel at the time.

To give the work of this panel a unified presentation—and because neither of our previous two reports had been printed in final, typeset volumes—

we have included our interim report and letter report in this volume as Parts II and III, respectively.

## 1–B  OVERVIEW OF THIS REPORT

This is our panel's third and final report; it is also somewhat different in tenor from the two reports that preceded it because—quite deliberately, and despite the formal name of our panel—the report has relatively little to do with suggesting changes or structures for the Census Bureau's 2010 CPEX program. We provide an updated description of what we know of the current plans for CPEX in Appendix B, because the Bureau has made some significant revisions to the program since the time of our letter report. But, although we critique the CPEX plans, we refrain from suggesting changes in this report for the simple reason that major change is effectively impossible due to the fast-approaching date of the 2010 census itself. At this writing, in late 2009, the operational exigencies of the decennial census and related programs are such that plans for experiments had to be finalized by spring 2009, if only to facilitate printing of questionnaires and selection of experimental samples. Hence, we do not offer specific comments or suggested refinements of the experiments to be conducted in 2010 because it is no longer feasible for such changes to be made.

Likewise, we reiterate our concern from previous reports about the acquisition of operational data from the 2010 census for evaluation purposes—it is impossible to speak meaningfully of the types of analyses that could be conducted in 2010 census evaluation studies without better knowledge of the operational data that will be available. But we must also accept that there is little to be gained by suggesting specific practices in this report because the Census Bureau's focus is now on getting its systems in order for the main count, a process that has grown more risky due to the need in 2008 to replan census operations and revert to paper-based nonresponse follow-up. We hope and trust that evaluation-ready data will be saved during the census process, but there are no further words at this point that can ensure that this is done.[1]

Instead, we turn in this final report to the broader themes raised by our charge and our work in the previous reports. Rather than the specifics of the 2010 CPEX, it is our intent in this report to speak of census *research and development* generally—to describe the role that research should play during the 2010–2020 decade to put the 2020 census on solid footing. In doing so, we also build on important contributions and themes raised by some of

---

[1] In late 2009, the Census Bureau described late changes to the CPEX program that include a "Master Trace Project" to retain and study 2010 census operational data. Because of this timing and because details of the CPEX changes are not yet available, we cannot comment further on these steps in the report. However, we describe them in introducing Section 3–A.2.

our predecessor National Research Council panels, including the Panel to Evaluate Alternative Census Methods (National Research Council, 1994), the Panel on Research on Future Census Methods ("2010 Census" panel; National Research Council, 2004b), and the Panel on Residence Rules in the Decennial Census (National Research Council, 2006).

In Chapter 2, we discuss what we think to be the most critical drivers for census planning and research: the balance between census cost and census quality. In Chapter 3, we offer a critique of the Census Bureau's current strategies for research, experimentation, and testing—in contrast with the Bureau's legacy of past research—before suggesting general directions for research in the 2010–2020 period. We also conclude that more formal, structural changes in the Census Bureau's organization for research are critical for success and elaborate those structural recommendations in Chapter 4. Two appendices to the report describe past and present Census Bureau research programs in detail; Appendix A describes the research, testing, and evaluation programs that preceded and were conducted in line with the 1950–2010 censuses, and Appendix B summarizes what we know about the 2010 CPEX program.

# – 2 –

# Planning the 2020 Census: Cost and Quality

L ESSONS AND CONCLUSIONS FROM THE HISTORY of the modern U.S. decennial census—and the role of research and development (R&D) in past decades—are vital to considering directions for effective R&D for the 2020 and subsequent censuses. Our reference to the historical record takes different shapes in the later chapters of this report, with Chapter 3 critiquing the Census Bureau's current research strategies with an eye on past efforts while Chapter 4 discusses organizational and structural features of the history of Census Bureau operational research.

This chapter looks at the historical record—particularly for the post–World War II decennial censuses—with a focus on broader forces. A lesson we learn from our historical review is that two key drivers—the costs of the decennial census and the quality of resulting census information—are the most important areas of concern for a successful census in 2020. Reconciling cost and quality involves trade-offs; the role of an R&D program such as we would like to see for the 2020 census is to provide the high-quality information and evidentiary basis for addressing those trade-offs. The history of the modern decennial census is also the story of a third set of factors—social and technological change—in which the census must operate and that offer both challenges and opportunities for a high-quality, cost-effective census in 2020.

We begin in Section 2–A with a general historical overview of the census in the post–World War II era to set the necessary context.[1]   We then turn in subsequent sections to examination of trends in census quality (2–B) and census costs (2–C). We close in Section 2–D with an assessment of how we think those two drivers should affect 2020 census planning in general.

## 2–A   DEVELOPMENT OF THE MODERN CENSUS

### 2–A.1   1940 and 1950: Sampling and R&D

The basic methodology of the 1940 and 1950 censuses was not dissimilar to that used in previous censuses: temporary Census Bureau employees (enumerators) went door to door, writing down answers provided by household respondents on large sheets of paper, or "schedules."[2]   Each schedule had one line per person for 30–40 people on the front and one line per housing unit on the back for the people listed on the front.[3]   The data were keypunched and tabulated by clerks using technology invented by Herman Hollerith for the 1890 census. Yet underlying the similarities were important innovations that paved the way for today's census.

The 1940 census was the first census to use newly developed probability sampling methods to ask a subset of the population some of the census content (6 of 60-odd questions); it was also the first census not only to include formal evaluations of the quality of the enumeration and of specific content items, but also to be followed by a program of pretests and experiments leading up to the next census (see Chapter 3 for details). This R&D program resulted in improvements to the wording of questions, enumerator instructions, and other features of the 1950 census.

The 1950 census included at least four important innovations. First, sampling was used much more extensively for collecting the census content: about two-fifths of the 60-odd questions were asked of samples of the population, which presumably contributed to the reduction in real dollar per housing unit costs in 1950 compared with 1940.[4]   Second, the first mainframe computer (UNIVAC I) for use outside academic and defense research

---

[1]Our synopsis of census-taking in 1940 through 2010 is based principally on Jenkins (2000) [1940 census]; Goldfield and Pemberton (2000a,b) [1950, 1960 censuses]; National Research Council (1985:Chap. 3, 5) [1970, 1980 censuses]; National Research Council (1995) [1990 census]; National Research Council (2004a) [1990, 2000 censuses]; National Research Council (2004b) [planning for the 2010 census].

[2]Enumerators were first hired in place of U.S. marshals for the 1880 census.

[3]The content of the census questionnaire expanded greatly from a few items in the first censuses to dozens of items in censuses of the late 19th century; in 1940, a census of housing was added to the census of population.

[4]The 1940 census was an outlier in the first half of the 20th century with regard to costs. It cost substantially more per housing unit than the 1930 census, so that the 1950 census costs reverted to the historical norm (see Section 2–C.1).

was delivered to the Census Bureau in spring 1951 in time to help process some of the census results. Third, the 1950 census included several experiments with far-reaching effects: tests of a household schedule in place of a line schedule, a housing-based sampling scheme for content instead of a person-based sampling scheme, self-enumeration in place of enumerator reporting, and variation of enumerator assignments in such a way as to be able to measure the error in content due to differences among enumerators. Fourth, the 1950 census included the first postenumeration survey to measure completeness of the census count.

### 2–A.2   1960: Mailout and the "Long Form"

The results of the enumerator variance experiments in the 1950 census, which indicated that the census was no more accurate than a 25 percent sample (Bailar, 2000), galvanized the Census Bureau to proceed with R&D on self-enumeration in place of personal visits as the dominant enumeration method (see Section A–1.b). The 1960 census was the first to use household questionnaires in place of line schedules and the first to use two different questionnaires—a "short form" with basic questions asked of every person and household and a "long form" with questions asked of a sample. (There were several variations of the long form with different sample sizes.) It was also the first census to mail out the questionnaires: shortly before Census Day, U.S. postal carriers dropped off unaddressed short forms to all housing units on their routes; households were instructed to fill out the forms and wait for an enumerator to pick them up and transcribe the responses to a computer-readable form. At every fourth household, the enumerator left one of the long forms to be completed by the household and mailed back to the census district office (in rural areas, enumerators completed the long form at the time of their visit).

The long-form response rate was 77 percent; enumerators revisited households that failed to mail back their long form to obtain their answers in person. A Bureau-invented device called FOSDIC (film optical sensing device for input to computers) was used to read microfilm images of the census questionnaires and transmit the data to mainframe computers for editing and tabulation. The effective use of computer technology presumably contributed to the modest reduction in real dollar per housing unit costs in 1960 compared with 1950.

### 2–A.3   1970–1980: Mailout-Mailback, Computerized Address List, Coverage Improvement, Dual-System Estimation

The 1970 census saw the implementation of mailout-mailback technology as we know it today, in which the Census Bureau develops a computerized address list, coded to census geographic areas (e.g., blocks, tracts, places, counties), uses postal carriers or census enumerators to deliver labeled questionnaires to every address on the list, and uses enumerators to follow up those addresses that do not mail back a questionnaire. The concept behind the development of an address list was to improve coverage and have control over each questionnaire rather than simply leaving it up to postal carriers and enumerators to do a complete job. To ensure that the new procedures would work well, the Census Bureau decided to limit their use to areas of the country containing 60 percent of the population; the remaining areas were enumerated in person.[5]

The 1980 census expanded mailout-mailback techniques to areas of the country containing 95 percent of the population. The 1980 census also greatly expanded coverage improvement efforts that were begun in 1970. While the completeness of the census count was a concern from the very first 1790 census, understanding of coverage errors in the census and their possible implications for the distribution of power and resources reached fever pitch in the years following the landmark "one-person, one-vote" decision of the U.S. Supreme Court in 1964 (*Baker v. Carr*). The Census Bureau adopted special coverage improvement programs for the 1970 census to obtain greater accuracy in the population counts because of their use for legislative redistricting and federal fund allocation and the belief that new methods were required to improve coverage given fears of being counted among some population groups, overlooked housing units in multiunit structures, and other factors. (By contrast, 1950 and 1960 census planning assumed that undercoverage was largely because enumerators failed to follow instructions—see U.S. Census Bureau, 1974:1.) The 1980 census greatly expanded and added to the coverage improvement programs used in 1970, spending about six times the amount spent in 1970 in real terms on such programs. For example, in 1980 enumerators rechecked units that appeared to be vacant or otherwise not eligible for the census on a 100 percent basis rather than for a small sample of the units as in 1970, and local officials were given the opportunity to review preliminary housing unit counts after nonresponse follow-up (Citro, 2000).

Finally, the 1980 census made the first use of dual-system methodology for estimating the net undercount and disparities in coverage for population groups by matching the results of an independent postenumeration survey to the census results in a sample of areas. (In contrast, the 1950 and 1960 postenumeration survey programs simply compared aggregate counts for the census and a postcensus recount by specially trained enumerators of a sam-

---

[5] Title 13 of the U.S. Code was modified in 1964 (P.L. 88-530) to eliminate the requirement that enumerators personally visit each dwelling; this change permitted use of mailout-mailback procedures in the 1970 and subsequent censuses.

ple of areas—this "do it again, better" method was shown to underestimate the undercount compared with dual-system estimation.) As discussed in Section 2–B, net coverage error declined to an all-time low in 1980 compared with the 1940–1970 censuses, but real dollar per housing unit costs of completing the census almost doubled compared with 1970, whereas the 1970 cost was about the same per housing unit as 1960.

### 2–A.4  1990–2000: Controversy Over Adjustment; Incremental Change

Ambitious plans were originally developed for both the 1990 and 2000 censuses to adopt the recommendations of many statisticians to use sampling for the count itself and not just the content in order to improve the completeness of census coverage and save on costs. Ultimately, the secretary of commerce decided not to adjust the 1990 census data for coverage error in July 1991. A court-ordered procedure to consider adjustment on the basis of a relatively small postenumeration survey ended with the courts upholding a decision by the secretary not to adjust the counts, even though measured net and differential undercount increased compared with 1980 (see Section 2–B).[6]

Similarly, plans to use sampling for nonresponse follow-up in the 2000 census were ruled to violate Title 13 by the U.S. Supreme Court in 1999, and problems with the large 2000 postenumeration survey led to widely supported decisions by the Census Bureau and the Department of Commerce not to adjust the census results for legislative redistricting or federal fund allocation. Indeed, the estimated net undercount for the 2000 census was close to zero, although this result reflected large numbers of duplicates almost offsetting equally large numbers of missed people.

Perhaps as a consequence of the attention devoted to the adjustment controversy and to refining the postenumeration survey methodology, neither the 1990 nor the 2000 census saw major innovations in census procedures of the magnitude of the use of computerized processing, mailout-mailback enumeration, and expanded coverage improvement programs that were introduced in the 1960–1980 censuses.[7] Essentially, the 1990 and 2000 censuses made incremental modifications to previous census procedures; they did not alter the paradigm of the modern census as established in 1980: development and checking of a computerized address list; mailout-mailback; in-person follow-up for nonresponse; coverage improvement operations;

---

[6]Adjusted counts at the national level by age, race, and sex were used to control the estimates from major household surveys in the decade of the 1990s, including the Current Population Survey, which is the source of official monthly unemployment and annual poverty statistics.

[7]However, the 2000 census was the first to capture both name and date of birth from completed census questionnaires, and to use those in matching studies to estimate duplicate census entries.

computerized editing and tabulation; and postcensus evaluation of coverage. Census net coverage worsened somewhat in 1990 and improved in 2000; census real dollar per housing unit costs increased by about 30 percent from 1980 to 1990 and by about 60 percent from 1990 to 2000. From 1960 to 2000, real dollar per housing unit costs increased by over 400 percent.

The experiments included in the 1990 and 2000 censuses were limited in scope (see Sections A–5.b and A–6.b) and did not set forth a clear path for innovation in 2010. An exception was the Census 2000 Supplementary Survey, which tested the ability to conduct a separate American Community Survey (ACS) to obtain long-form questionnaire content at the same time as a full census (see next section). An experiment in using administrative records to substitute for a traditional census or for nonresponse follow-up was well conducted but, because of a late start in planning and limited resources, was limited in scope (Bye and Judson, 2004:1–2). The evaluations of census procedures in 1990 and especially 2000 were limited in usefulness for future census planning, consisting largely of descriptive reports that documented the inputs and outputs of particular procedures but did not provide rigorous cost-benefit analysis of them or an assessment of their relative effectiveness for different kinds of geographic areas or population groups.

## 2–A.5   2010: Recovery from Near Disaster?

By 2001, the Census Bureau articulated a strategy for a "reengineered" census process in 2010; it also announced that adjustment for measured net undercount would be off the table for 2010. Save for the pilot work that had been done on the ACS, this emergent strategy for 2010 did not extend directly from the 2000 census evaluations and experiments for the simple reason that it could not do so chronologically—most of the evaluation reports were only completed and released in 2002 and 2003. However, findings from the 2000 census experience would later influence 2010 census plans—for instance, in the decision to add two coverage probe questions to the 2010 census questionnaire.

The reengineered 2010 process hinged critically on three major initiatives. First, the Census Bureau planned to modernize its Topologically Integrated Geographic Encoding and Referencing (TIGER) system—the geographic database used to map census addresses and code them to specific census blocks (and thus to higher-level aggregates like cities or school districts). When it was developed in the 1980s, TIGER represented a significant improvement over the patchwork address coding guides that were used in the 1970 and 1980 censuses, but after almost 20 years of use, it needed realignment of its geographic data (through comparison with local geographic information system files) and overhaul of its software structure. The Census Bureau embarked on a multiyear Master Address File (MAF)/TIGER Enhancements Program as a key plank in its 2010 census plan; the major component of this program was a contract (the MAF/TIGER Accuracy Improvement Program) to perform the complete realignment of TIGER features in electronic files, which was awarded to Harris Corporation and carried out.

Second, the Census Bureau committed to replacing the census long-form sample—a detailed battery of social and economic questions administered to samples of census respondents—with the continuous American Community Survey. The idea for conducting a "rolling census" goes back several decades; formal planning of a continuous ACS began in the early 1990s, and pilot data collection began in 1996 in 4 counties, later expanded to about 30 counties. The Census 2000 Supplementary Survey, conducted in about one-third of all counties, confirmed that the Bureau could successfully field the ACS and the decennial census at the same time. This larger-scale administration also yielded data that could be compared with the 2000 long-form sample to assess the adequacy of the new survey. Based on the results, the Bureau decided that the 2010 and subsequent censuses would include only the short-form items and that the ACS would go into full production as soon as funding became available, which occurred in 2005.

Third, the Census Bureau decided to use modern handheld computer technology for two key census processes: checking the address list in 2009 and conducting nonresponse follow-up enumeration in 2010. The use of such technology was expected to significantly reduce the amount of paper (questionnaires, enumerator timesheets, maps, etc.) and office space required for the census, permit real-time monitoring of census field operations, and reduce census costs compared with paper-and-pencil methods. However, the R&D program for the handheld technology and the contract for implementation, also with the Harris Corporation, were poorly planned and executed, necessitating an extensive "replan" effort in early 2008 (see Section 2–C.2), which resulted in the decision to revert to paper-and-pencil methods for nonresponse follow-up operations.

Entering the 2010 planning cycle, the Census Bureau hoped that its 2008 "dress rehearsal" would be exactly that—a full operational pretest. The Bureau's 1998 dress rehearsal for the 2000 census was less a rehearsal than a major experimental comparison of three competing census designs, each making different use of sampling and coverage measurement. However, the 2008 dress rehearsal in San Joaquin County, CA, and the Fayetteville, NC, area was not able to function as a full dress rehearsal. The planned rehearsal was already scaled back because of the budgetary constraints of operating for key periods under continuing resolution, previous-fiscal-year-level levels. But, with the early 2008 replan, the Bureau could not conduct a nonresponse follow-up operation in its dress rehearsal for the basic reason that the late reversion to paper-based methods left it without a nonresponse follow-up operation to test. The lack of a full-fledged dress rehearsal left the 2010 census with many unanswered questions as to its procedures and plans.

We discuss the implications of these plans and developments on estimated 2010 census costs in Section 2–C. Of course, the effects on completeness of coverage in the 2010 census will not be known until after the census and its associated coverage measurement programs are conducted.

## 2–B   CENSUS QUALITY

Perhaps the most critical driver of decennial census planning and execution is the concern that the census achieve as complete coverage of the population as possible, including not only the total number of inhabitants, but also their distribution by state and other geographic areas and their racial and ethnic composition. This concern dates back to the first census in 1790, when Secretary of State Thomas Jefferson expressed the view that the census count was short of the "true" population, which he believed to be 4.1 million people and not 3.9 million as the census reported (Wells, 2000:116). Subsequent censuses also raised concerns about coverage—notably, in 1870, complaints of undercounts in New York City and Philadelphia led President Ulysses S. Grant to order a recount, which, however, added only 2 percent and 2.5 percent, respectively, to the two cities' population totals. The dramatic growth in the population of the South between 1870 and 1880 ultimately led the 1890 census office to estimate that the 1870 census had undercounted the South by 10 percent and the country as a whole by 3 percent (Hacker, 2000b:129).

In the 20th century, the findings from the 1920 census that the population was more urban than rural for the first time in the country's history led to attacks on the accuracy of the census, and, for the first time, Congress was not able to agree on a reapportionment of the House of Representatives to reflect the census results in a timely manner (McMillen, 2000:37–38). Following the civil rights revolution and the "one person, one vote" Supreme Court decisions, concern about undercount fueled controversy over whether to adjust the census for measured net undercount and to correct disparities in coverage by geographic area and population group and drove the planning, execution, and evaluation of the 1980–2000 censuses.

### 2–B.1   Definition and Measures

We use the term "census quality" to denote the accuracy of the count and its basic distribution by geography and population group. Since formal coverage evaluation began for selected population groups in the 1940 census and for the population as a whole in 1950, two general quality metrics have dominated the discussion. The first metric is net coverage error, simply the difference between census-based counts or estimates and their associated true values for some geographic or demographic domain; estimated counts greater than the true values are then dubbed a net overcoverage, while estimates lower than the true value represent net undercoverage. Because the true values are unknown, competing strategies have been devised to derive approximations—typically through an independent effort to estimate the same population (postenumeration survey) or derivation of estimates based on birth, death, and migration data (demographic analysis). To provide comparability across domains, net coverage error is often expressed as a percentage of the "true" counts. Due to the historical undercoverage of racial and ethnic minority groups, the second type of quality metric— differential net undercoverage—focuses on the difference between the rate of net undercoverage error for a given demographic group compared with the national rate.

Census undercoverage and overcoverage are made up of two general types of errors:

- *omission*, which occurs when a current resident of the United States is not included in the census anywhere, and

- *erroneous enumeration*, which occurs when a nonresident (or nonperson) is erroneously included anywhere in the census or when a person is included more than once.

Coverage error may also arise when a (nonduplicated) resident is counted at the incorrect geographic location. The severity of this latter type of error depends on the degree of geographic displacement and the level of aggregation of interest; depending on those perspectives, geographic misallocations are either moot (e.g., when a person in one block is listed in the wrong block, but both blocks in question are within the same county for which coverage is being estimated) or count as two errors (undercoverage in one block and overcoverage in another).

### 2–B.2    Net Coverage Error, 1940–2000

Table 2-1 shows estimated net coverage error, by the method of demographic analysis, and the difference in coverage estimates for blacks and all others, for the 1940 through 2000 censuses. With the exception of 1990, when the net undercoverage rate increased from the previous census, there has been a sustained trend toward more complete coverage of the total population. Whereas the 1940 census had an estimated net undercount rate as high as 5.4 percent, the 2000 census achieved an estimated net undercount rate of practically zero (0.1 percent). Estimated net undercount rates for blacks and nonblacks also declined over the period (with the exception of the uptick in 1990 for both groups), although the difference between black

**Table 2-1** Estimates of Percentage Net Undercount, by Race, from Demographic Analysis, 1940–2000 (in Percent)

|  | 1940 | 1950 | 1960 | 1970 | 1980 | 1990 | 2000 |
|---|---|---|---|---|---|---|---|
| Total population | 5.4 | 4.1 | 3.1 | 2.7 | 1.2 | 1.7 | 0.1 |
| Black | 8.4 | 7.5 | 6.6 | 6.5 | 4.5 | 5.5 | 2.8 |
| Nonblack | 5.0 | 3.8 | 2.7 | 2.2 | 0.8 | 1.1 | −0.3 |
| Difference (percentage points) | 3.4 | 3.6 | 3.9 | 4.3 | 3.7 | 4.4 | 3.1 |

NOTE: Minus sign (−) indicates net overcount.

SOURCES: 1940–1980: National Research Council (1995:Table 2.1); 1990–2000: National Research Council (2004a:Table 5.3, columns for "revised (October 2001)").

and nonblack net undercount rates increased from 3.4 percent in 1940 to 4.3 percent in 1970, and was as high as 4.4 percent in 1990, before declining to 3.1 percent in 2000.

Demographic analysis does not provide estimates of coverage error for other population groups, such as Hispanics, but coverage rates for those groups appear to have improved as well. Thus, based on postenumeration survey methodology, net undercount decreased from 5 percent in 1990 to 0.7 percent in 2000 for Hispanics, from 4.6 percent in 1990 to 1.8 percent in 2000 for non-Hispanic blacks, from 2.4 percent in 1990 to a net overcount of 0.8 percent in 2000 for non-Hispanic Asians, and from 0.7 percent in 1990 to a net overcount of 1.1 percent in 2000 for non-Hispanic white and other races (National Research Council, 2004a:Table 6.7).

## 2–B.3 Another Metric: Gross Coverage Errors

Undoubtedly, the achievements in reducing estimated net undercount and narrowing the differences between estimated net undercount rates for racial and ethnic groups over the 1940–2000 period are due, in no small measure, to the proactive efforts by the Census Bureau, in cooperation with many public- and private-sector organizations, to improve coverage. As noted above, the addition of coverage improvement operations to the conduct of the census began in 1970 and greatly escalated in 1980.

Yet research has shown that coverage improvement programs not only add people to the census who may have been missed otherwise, but also add people who should not be counted at all or who may have been counted elsewhere. Examples include the 1980 Vacant/Delete Check program (U.S. Census Bureau, 1989:Ch. 8) and the 1990 program to count parolees and probationers (Ericksen et al., 1991:43–46). In 2000, the problem-plagued development of the MAF from multiple sources, some not used in previous censuses, contributed to large numbers of duplicate enumerations, only some of which were weeded out in subsequent census operations (National Research Council, 2004a:142–143).

Estimates of gross errors, including both erroneous enumerations and omissions, are as large as 36.6 million in 1990 (16.3 million erroneous enumerations and 20.3 million omissions) and 33.1 million in 2000 (17.2 million erroneous enumerations and 15.9 million omissions) (National Research Council, 2004a:253). Too much should not be made of the specific numbers, given that some errors are not of consequence for larger areas of geography and given different definitions and methods for estimating gross errors in the two censuses, but, however defined, the census is and has always been far from error-free.

## 2–C    CENSUS COSTS

It appears throughout much of the history of the U.S. census—and particularly in the period after 1970—that concerns about coverage have trumped concerns about costs, with Congress willing to appropriate ample funds for the conduct of the census. However, cost increases from census to census are not written into stone—in fact, real dollar per person or housing unit costs have held steady and even declined in some censuses, as seen in the next section. Yet in the period 1970 to 2010, costs have escalated enormously, and the increases appear harder and harder to justify. Because of this continued growth in census costs, it seems highly likely that containing costs—while maintaining or improving census quality—will and should be a major driver of 2020 census planning.

In the discussion that follows of historical census costs, it is important to note the difficulties in obtaining comparable cost estimates across time. Comparisons can be affected by the choice of a specific price deflator; also, it is not clear that what is included in census costs is strictly comparable from census to census. Thus, the reader should consider the data provided as indicative of the order of magnitude of the costs from one census to the next. Following usual practice, we discuss "life-cycle" costs, which are inclusive of precensus planning and testing, the actual conduct of the census, and processing and dissemination of census results. The limitations of available census cost data make it difficult to decompose cost increases (or decreases) by examining specific operations.[8]    In fact, the absence to date of a robust, comprehensible, fully parameterized cost model for the census and its components makes it difficult not only to analyze the reasons for changes in costs

---

[8]Another way of describing the situation is that the Census Bureau's cost accounting is not program-oriented; costs for broad categories such as office space may be known, but not how those costs may be disaggregated by specific programs or operations.

between censuses, but also to plan for the next census. Moreover, with few exceptions, there has been little analysis of the cost-benefit ratio, in terms of coverage improvement, of various census operations.

## 2–C.1   Census Costs Over Time

Since the first census, the American population has increased manyfold (from 3.9 million people in 1790 to over 300 million people in 2009) and become more diverse in living arrangements and many other ways; the needs of Congress and the executive branch for additional information beyond a basic head count have also grown. Hence, it is not surprising that the costs and complexity of the census have increased commensurately. There were only about 650 enumerators (U.S. marshals) for the 1790 census, whereas recent censuses have employed half a million or more enumerators plus postal service carriers; the office force for the census has also increased over time, as has the volume of data collected and produced for public use. By 1900, the real dollar per person cost of the census had increased from 11 cents to $3 in 1999 dollars. For the 1900–1960 period, however, despite continued population growth, costs remained in the range of $3–4 most of the time. Exceptions occurred for the 1920 census, which cost about 30 percent less than the average, and the 1940 census, which cost about 60 percent more than the average (Anderson, 2000:384—estimates are not available per housing unit).

Turning to census costs from 1960 to the present, we focus on the magnitude of estimates of real dollar costs per housing unit, which is the appropriate measure for the modern mail census. Table 2-2 shows the estimated per housing unit cost for each census beginning in 1960 in both nominal and real 2009 dollars. In real terms, the 1970 census cost only slightly more than the 1960 census, but costs per household increased by over 90 percent from 1970 to 1980—the single largest percentage increase in the costs of census-taking since the Census Bureau was established as a permanent agency in 1902. The increase from 1980 to 1990 was only 30 percent, while the increase from 1990 to 2000 was over 60 percent. Although it is not and cannot be known at this writing what the full costs of the 2010 census will be, current estimates suggest that they could amount to $115 per housing unit (including the ACS and the TIGER modernization program), an increase of 64 percent over 2000 and a cumulative increase of more than 600 percent over 1960 in real dollar terms.

A major factor in the substantial cost increase between the 1970 and 1980 censuses was that the enumerator workforce more than doubled in size, as did headquarters and processing staff. District field offices also stayed open several months longer on average (6–9 months) in 1980 compared with 1970 (4–6 months) and 1960 (3–6 months) (National Research Coun-

**Table 2-2**  Per Housing Unit Costs, 1960–2010 U.S. Censuses

|  | 1960 | 1970 | 1980 | 1990 | 2000 | 2010[a] |
|---|---|---|---|---|---|---|
| Full-cycle census costs |  |  |  |  |  |  |
| (millions, nominal dollars) | $120 | $231 | $1,136 | $2,600 | $6,600 | $14,700 |
| Population (millions) | 179.3 | 203.3 | 226.5 | 248.7 | 281.4 | 301.6 |
| Housing units (millions) | 58.9 | 70.7 | 90.1 | 104.0 | 115.9 | 127.9 |
| *Costs in 2009 real dollars* |  |  |  |  |  |  |
| Total (millions) | $995 | $1,206 | $2,947 | $4,424 | $8,060 | $14,700 |
| Per housing unit | $16.89 | $17.06 | $32.71 | $42.54 | $69.54 | $114.93 |

NOTES: 2010 estimates of population and housing units are the U.S. Census Bureau's 2007-vintage estimates. For the cost of the 2010 census, we have used the high end of the Census Bureau's budget submission for fiscal year 2009 (U.S. Census Bureau, 2009b:CEN-154), which includes the cost of the MAF/TIGER Enhancements Program and the American Community Survey. Nominal dollars are converted to real 2009 dollars by the gross domestic product chain-type price indexes for federal government nondefense consumption expenditures (based on Table 3.10.4, line 34, at http://www.bea.gov/national/nipaweb/SelectTable.asp?Selected=N\#S3 [5/25/09]. Based on the life-cycle cost estimates shown in Table 2-3, the Census Bureau estimated the per housing unit (excluding group quarters) cost for the 2010 census as $93.58 in constant 2010 dollars; see text for discussion of the Table 2-3 figures.

[a] Estimated, 2009 dollars.

SOURCES: National Research Council (1995:Table 3.1); National Research Council (2004a:Table 3.1).

cil, 1995:Table 3.3). One reason for the scaling up of field operations was the greatly expanded effort in 1980, referenced above, to reduce under-count by implementing a variety of coverage improvement programs, such as rechecks of the address list, a 100 percent Vacant/Delete Check, and oth-ers. Tables 2-4 to 2-8 illustrate the increase in the number and extent of such operations by listing the various coverage improvement methods used in the 1970–2000 censuses and planned for the 2010 census; because of the length of these tables, they are placed at the end of the chapter and numbered accordingly.

The real dollar per housing unit cost increase between 1980 and 1990 was only a third as much as that between 1970 and 1980. Some of the same factors contributed to increases as in 1980: offices stayed open even longer (9–12 months), and there were continued, although less marked, increases in staff. An analysis of the cumulative cost increases between 1970 and 1990 questioned the effectiveness of about two-thirds (over $1 billion in 2009 dollars) of the added real dollar per housing unit costs, even after taking account of a substantial decline in the mail response rate (from 78 percent in 1980 to 75 percent in 1980 to 65 percent in 1990), which necessitated many more expensive enumerator visits to households to obtain their data (National Research Council, 1995:Table 3.2, 50–55).

The 2000 census incurred major additional expenses due to the need to plan for two different censuses, one incorporating sampling for nonresponse follow-up and adjustment of the census counts and the other a "traditional" census, which was demanded by the congressional majority party in the second half of the 1990s. The final census plan was not confirmed until one year prior to Census Day, following the February 1999 U.S. Supreme Court decision that sampling as part of the census enumeration violated census law for apportionment counts. Recognizing the difficulties for planning the 2000 census caused by the political conflict between Congress and the administration about an appropriate census design, the appropriations for 2000 were greatly increased to enable the Census Bureau to hire the necessary staff to throw into the effort—for example, to rewrite software programs that had been developed assuming one design and subsequently had to be revamped for the final design. The 2000 census stemmed the decline in the mail response rate, which was an important achievement and also, as noted earlier, produced a close to zero net undercount.

### 2–C.2 Life-Cycle Cost of the 2010 Census

The latest projections of the anticipated costs of the 2010 census make clear that it will be the most expensive in the nation's history, representing a significant increase in per housing unit costs over the 2000 census. We now review the history of evolving cost estimates of the 2010 census.

In order to obtain funding for all components of the 2010 census reengineering, the Census Bureau produced an initial document on projected life-cycle costs in June 2001 (U.S. Census Bureau, 2001). This document was revised periodically; a September 2005 version showed a total estimated cost for the 2010 census of $11.255 billion—$1.707 billion for the ACS, $228.9 million for the MAF/TIGER Enhancements Program, and $9.013 billion for the core, short-form-only census program (U.S. Census Bureau, 2005:3).

The Census Bureau's budget submission in 2007 to Congress for fiscal year 2008 (U.S. Census Bureau, 2007:CEN-159) noted:

> In June 2001, the Census Bureau estimated that the life cycle costs of a 2010 Decennial Census Program that repeated the Census 2000 approach would be $11.725 billion, while the estimated lifecycle cost for the reengineered design was estimated to be $11.280 billion—a savings of $445 million. [The] estimated life cycle cost for the 2010 Decennial Census Program now stands at $11.525 billion [an increase of $245 million in 2009 dollars from the 2001 estimate]. However, the forecasted savings from pursuing the re-engineered design now are estimated to be $1.409 billion because the estimated life cycle cost of reverting now

to a Census 2000 design is $12.934 billion [an increase from the 2001 estimate of $1.209 billion].

At that point, in 2007, the Census Bureau also expressed confidence that the 2010 census would "enjoy the lowest rate of cost increase in the last four decades." Over those previous censuses, the average rate of cost increase, according to the Bureau, was 70.7 percent, and a basic extrapolation based on that growth rate would price the 2010 census at $15.1 billion or about $116.42 per household—substantially higher than the figures the Bureau projected for its reengineered process (U.S. Census Bureau, 2007:CEN-161). The Census Bureau reiterated the same discussion in its congressional budget submission for fiscal year 2009 (issued in February 2008), although it acknowledged a $20 million increase in projected life-cycle costs mainly due to changes in assumptions regarding mileage costs, office leasing costs, and number of housing units (U.S. Census Bureau, 2008a:CEN-166–CEN-167).

We see a number of problems with the Census Bureau's budget submissions for 2010. The first is that the February 2008 estimate did not acknowledge the likelihood of cost increases due to the problematic performance of the handheld technology, which was known inside the Bureau but not yet outside it. The second is that there was no explanation of the changing estimates for conducting a traditional census in 2010 between the 2001 and 2007 estimates. Even more important, there has been no acknowledgement or awareness by the Bureau that forecasting increased costs for the traditional census at a rate of 70.7 percent might not be a reasonable or prudent thing to do. The historical cost data, even with their limitations, make it clear that the biggest increase in census costs occurred between 1970 and 1980 and also that the next biggest increase—that between 1990 and 2000—was due largely to the problems of being forced to plan for and test two different designs for the census. We see no justifiable basis for projecting traditional census costs as a straight line extrapolation of the average increase over the past few decades.

We requested from the Census Bureau its current estimated life-cycle costs for the major activities of the 2010 census (for comparative purposes, we also requested comparable category costs for the 2000 census but were not provided with those data). The provided life-cycle costs (provided in June 2009) are shown in Table 2-3. It is apparent that the estimates presented in the table refer strictly to the short-form-only census and not the entire 2010 census planning and implementation cycle: adding the MAF/TIGER Enhancements Program and the ACS by using the previously cited estimates of $239 million and $1.707 million, respectively, would bring the bottom line in Table 2-2 to $14.5 billion in nominal (2009) dollars, or about $3 billion higher than the estimates of $11.5 billion cited above.

**Table 2-3**   Census Bureau Life-Cycle Cost Estimates for the 2010
Census (covering fiscal years 2002–2013)

| Key Activity | Life-Cycle Cost (millions of dollars) |
|---|---|
| *Major Contracts* | |
| Decennial Response Integration System (DRIS) | 981 |
| Field Data Collection Automation (FDCA) | 801 |
| Communications | 308 |
| Data Access and Dissemination System (DADS) | 176 |
| Printing | 179 |
| Mailout-Mailback Postage | 257 |
| *Office Space and Staff* | |
| Regional Census Centers (RCCs) | 828 |
| Local Census Offices (LCOs) | 1,301 |
| *Field Operations* | |
| Address Canvassing | 386 |
| Group Quarters Advance Visit | 17 |
| Group Quarters Enumeration | 80 |
| Group Quarters Validation | 71 |
| Coverage Measurement | 83 |
| Puerto Rico | 62 |
| Island Areas MOAs | 37 |
| Field Verification | 39 |
| Nonresponse Follow-Up | 2,744 |
| Vacant/Delete | 341 |
| Military | 5 |
| Remote Alaska | 4 |
| Service-Based Enumeration | 41 |
| Transient Night | 11 |
| Update Enumerate | 108 |
| Update Leave | 116 |
| Urban Update Leave | 2 |
| Fingerprinting | 148 |
| National Processing Center Census Operations | 364 |
| Headquarters Staff and All Other | 2,986 |
| **Life-Cycle Total Costs in Nominal Dollars** | **12,522** |

NOTE: Excludes costs for the American Community Survey and the MAF/TIGER
Enhancements Program; see text.
SOURCE: Estimates provided to the panel by the U.S. Census Bureau, June 4, 2009.

The added costs for 2010 are due largely to the need to "replan" crucial
census operations in the wake of problems with the Bureau's plans for using handheld computers in the 2010 census. In spring and summer 2007,
the handhelds received their first full field test through their use in the address canvassing operation for the 2008 census dress rehearsal. Problems
experienced with the devices during the operations prompted the Bureau to
commission a review of its Field Data Collection Automation (FDCA) con-

tract with Harris Corporation, a contract involving not only the handheld computers, but also the development of various operations control systems (OCSs) that govern the flow of information during the census, as well as installation of computer equipment in field offices. This review prompted the Census Bureau to formalize a set of requirements for the FDCA contract and, in turn, for Harris to provide a "rough order of magnitude" estimate of additional funds needed to meet those requirements. The size of that estimate led then–newly on the job Census Director Steve Murdock in February 2008 to establish an internal task force to evaluate options for the FDCA contract; that task force recommended as its "replan" option that Harris retain authority for developing handheld computers for address canvassing but that the Census Bureau assume authority for a paper-based nonresponse follow-up operation. Harris was also initially slated to retain full authority for the OCS development, but the responsibility for OCSs other than that for the address canvassing was assumed by the Bureau later in 2008.

On April 4, 2008, Commerce Secretary Carlos Gutierrez testified to House appropriators that:[9]

> The effect of moving forward with this alternative, as well as the non-FDCA related planning challenges we have faced will require an increase of $2.2 to $3 billion dollars through Fiscal Year (FY) 2013. This will bring the total lifecycle cost of the 2010 Census to between $13.7 to $14.5 billion.
>
> These costs are driven in large part by increases in the numbers of people who will be needed to carry out the 2010 Census; these include enumerators and personnel to service the help desks, data centers, and the control system for the paper-based [nonresponse follow-up (NRFU)]. There are also additional costs that result from more recent increases in gas prices, postage, and printing.

To this end, the Census Bureau requested additional funding in fiscal year 2008 (largely through transfers from other Commerce Department programs) and formally submitted an amendment to its fiscal year 2009 budget request seeking an additional $546 million (U.S. Census Bureau, 2008b). In addition, the American Recovery and Reinvestment Act of 2009 (P.L. 111-5) economic stimulus package included provision of $1 billion for the 2010 census (made as a direct addition to the Periodic Censuses and Programs account), with explanatory text that one-quarter of the funds is intended for use in partnership and outreach programs. On the basis of these changes, the Census Bureau's budget submission to Congress for fiscal year 2010 (dated May 2009) asserts that (U.S. Census Bureau, 2009b:CEN-154):

> After factoring in appropriations for FY 2002 through FY 2008, the President's Budget request for FY 2009, and ongoing programmatic en-

---

[9]The secretary's prepared testimony is available at http://www.commerce.gov/NewsRoom/SecretarySpeeches/PROD01_005468.

hancements or changes due to new requirements, the estimated life cycle cost for the 2010 Decennial Census Program now stands at $14.7
billion (in nominal dollars). The life cycle estimate has been revised to
reflect the Field Data Collection Automation Program rescope, including the Census Bureau's assuming activities descoped from that contract
and increases to a contingency fund based on increased risk. Additional
changes include: (1) higher estimated mileage costs (increased to 62.5
cents/mile) that we will have to pay to over 1,000,000 temporary employees in FY 2010; and (2) higher costs for Census Coverage Measurement field activities due to revised work hour assumptions.

We understand the difficulties in costing out such an extensive operation
as the decennial census, but the Census Bureau should be able to provide
information on component operations that is more useful for decomposing
the factors that led to the 2010 cost estimates prior to and after the problems
with the handheld contract. It is also uncertain why comparable figures for
the 2000 census are not more forthcoming. Still, the rough level of detail in
Table 2-3 does make clear one basic truth about the cost of census-taking:
that the NRFU operation, and the assumptions made about its conduct, is a
critical driver of overall census costs. The table also speaks to the complexity
of the census as it is currently conducted in terms of staff and space needs;
the entries for office space and staff at Census Bureau headquarters and the
regional and local census offices constitute over 40 percent of the total costs.

## 2–C.3  Comparison with Other Censuses

Systematic, time-series information on the lifecycle costs of censuses in
countries that conduct a traditional-type census (as opposed to relying on a
population register, rolling sample, or other means to determine population
counts) is difficult to obtain. The United Nations Statistics Division conducted a survey of national census offices on expected "total cost of population and housing census" for the "2010 round" of censuses in participating
countries, but the survey did not ask for comparable information for the
2000 round, nor does the report on the survey data provide detail on major
components of those expected costs (Stukel, 2008).

The snapshots of costs that are available suggest that other countries have
also experienced increases in per household or per capita costs in recent census rounds. For example, the United Kingdom Cabinet Office has projected
the 2005–2016 costs for the 2011 census in England and Wales at £481.7
million (Cabinet Office, 2008:1.28); the 2001 census was said to have cost
on the order of £260 million and the 1991 census £140 million (Geographical, 2004). Other countries have experienced sufficient growth in census
costs as to take cost reduction as a first precept of census planning; the Australian Bureau of Statistics (ABS) commenced planning for its 2006 census
"on the basis that the real per capita cost of conducting the 2006 Census be

no more than the cost of conducting the 2001 Census" (Trewin, 2006:5). In still other cases, census costs continue to grow but in a somewhat more contained way than the U.S. experience. The estimated cost per household of the census in the Republic of Korea increased 116 percent from 1985 to 1990 and 125 percent again from 1990 to 1995; thereafter, Korean census officials have succeeded in bringing the rate of growth down to just over 39 percent for 1995–2000 and 2000–2005 (with the average per household cost in 2005 being 8,065 Won or about U.S. $6.5 dollars; National Statistical Office, Republic of Korea, 2006:Table 1).

A useful example is that of Canada, in which fixed costs (adjusted for population growth) are assumed by census planners and built into census designs. In real 2008 Canadian dollars, the per housing unit cost of recent quinquennial censuses in Canada has held stable. In particular, the per dwelling cost of the 1996 Canadian census (in 2008 dollars) was $38.85, growing only to $40.32 in 2006; Statistics Canada currently estimates that the per dwelling cost for its 2011 census will decrease to $39.98.[10]

To be sure, censuses in other countries are not strictly comparable to the U.S. experience; the U.S. census presents significant challenges in terms of diversity, size, and distribution of the population and in trust in and receptiveness to government programs. That said, as we discuss in Section 3–A.2, the argument of the "special nature" of the U.S. census can be carried too far.

### 2–C.4    Coverage and Costs

Our historical review has found that census costs will have escalated by more than 600 percent over the period 1960–2010, even after adjusting for inflation and the growth in housing units. We have suggested that a major factor in cost escalation is the efforts by the Census Bureau, supported by Congress and the executive branch, to reduce coverage error to the greatest extent possible. Recent censuses have introduced new and seemingly better coverage improvement operations, layering operations to try to get more and better information on specific difficult-to-count population groups: college students, people who move on or around Census Day, people who live in housing projects, parolees and probationers, and so forth. Although these coverage improvement programs have value, they are not cost-free. The consequence for census cost has been a steady accretion of coverage improvement programs and other procedures that—once added—become difficult to subtract, lest census coverage appear to be harmed.

By 2000, the Census Bureau had achieved a major success in reducing levels of net undercoverage, to the point of yielding an estimated zero net

---

[10]Figures compiled from Statistics Canada data by panel member Ivan Fellegi.

national undercount, the first in census history. But the focus on net coverage masks significant levels of gross census errors—omissions, duplicate or erroneous enumerations, and geographic displacement—that fell into a delicate near-balance when considered in national net. The results of 2000 suggest the importance of studying the nature and extent of gross census errors and further scrutinizing the components of census error. Indeed, they suggest that the census-taking in the United States has reached the point at which the long-standing goal of reducing net undercount is no longer quite apt—that the steady accretion of coverage-building operations needs to be balanced with operations for detecting and filtering duplicates and diagnosis of the unique contributions (and gaps) in each step of increasing coverage (e.g., adding addresses and making questionnaires more widely available). Without research and careful inquiry into the components of error and the contributions of individual operations, the census is arguably at the point at which introducing further complexity to the enumeration process in an effort to reduce net undercoverage could conceivably add more error than it removes as well as adding costs.

## 2–D   ASSESSMENT: TIME TO RETHINK THE CENSUS AND CENSUS RESEARCH

If one accepts the premise that cost and quality are two critical factors in the decennial census, then we think that the preceding discussion makes clear the key point that we wish to express in this report: effective planning for the 2020 census must reflect the concept that the cost of conducting the census has grown out of control in recent decades and that increased spending—alone, and applied unwisely—is unlikely to radically increase the quality of the census. In beginning to conceptualize the 2020 census, we think it appropriate to take as a precept that an incremental approach to 2020 census planning is simply untenable. Simply assuming the life-cycle costs of the 2010 census as a funding base and scaling up from there—thus arriving at a 2020 census that costs in excess of $20 billion (in 2010 dollars) but uses methodology not substantially different from the 1970 census—is unworkable.

To be sure, the most recent censuses have included significant new additions to methodology and enumeration procedures. That said, the 2010 census follows the same basic fundamentals as the 1970 census, relying heavily on mailout-mailback methods (and, with it, reliance on an address register to associate census responses from households with address locations), coupled with programs to supplement census coverage in areas not amenable to mail. There are a variety of reasons that should compel attention to the basic assumptions of census-taking and, ideally, lead to a 2020 census that is significantly different from its predecessors. These reasons include:

- *Continued complexity of household arrangements and ties to geographic place:* As described in more detail by the National Research Council (2006) Panel on Residence Rules in the Decennial Census, segments of the population are becoming more difficult to accurately enumerate using traditional methodologies. Broad societal trends have made it increasingly difficult to uniquely identify the single "usual residence" envisioned by the census; these situations include children of divorced or separated parents in joint custody situations, long-distance commuting patterns and "commuter marriages" in which spouses work in different locations, prisoner reentry into the community in the wake of the 1980s and 1990s surge in correctional populations, and seasonal migration based on extreme summer or winter temperatures. Likewise, recent experience with shifting economic conditions (high rates of home foreclosure) and natural disasters (displacement of Gulf Coast residents by Hurricanes Katrina and Rita in 2005) have suggested other major enumeration difficulties.

- *Decreased cooperation with surveys and confidentiality concerns:* Declining response rates have been a major concern in survey research in recent years, owing in part to factors such as increased reliance on mobile phones (in some cases, to the exclusion of household land lines) and use of caller ID services to screen calls. Thus far, the Census Bureau has been able to hold off major dips in response in its demographic surveys, and the decennial census and the ACS are aided in this regard by their mandatory-by-law nature.

- *Technological advances:* As we discuss in this and our earlier reports, the Census Bureau has recently had a mixed record with regard to technological advances currently available that offer opportunities for facilitating various aspects of census-taking. The Internet is now an ubiquitous part of society, and it offers a number of ways for easing data collection and increasing data quality. Although the 2000 census made the United States one of the first countries to offer an Internet response option to a census, the Census Bureau declined to permit Internet response in 2010 and has no current plans to use the Internet on even an experimental basis in 2010—even as other countries have turned to Internet response to meet public expectations and possibly reduce some data-processing costs. Handheld computing devices are also now ubiquitous, with smartphones serving as full-fledged computers and portable devices being commonly used by delivery services and other service-oriented businesses to automate paper-based activities. The Census Bureau approached the 2000–2010 decade with a vision of using such handheld devices for major census operations, but multiple factors (including a failure to develop requirements in a timely fashion) forced a costly and risky late-stage switch back to paper-based nonresponse follow-up.

- *Increased availability and quality of administrative data:* Administrative records databases (including those maintained by government programs as well as those in the private sector) have grown in their coverage and completeness. Although administrative records were a focus of one of the formal experiments of the 2000 census, the Census Bureau has not yet integrated records databases into all steps of the census process to the extent that is possible. Thorough examination of the uses of administrative data in the census context will require working with Congress and possibly amending Title 13 of the U.S. Code, not only to secure access and permissions to conduct research using the data but also to resolve the question of whether reference to administrative data (in part or in whole) can constitute a census enumeration.

- *Future of mail service:* Related to the technological advances point is another reason why core reliance on the mailout-mailback model may require attention in coming years: the coming decade is likely to be a pivotal one for the U.S. Postal Service (USPS) as well, as it grapples with sharp declines in the volume of total mail. Since 2007, USPS has recorded annual losses of $5 billion or more and currently projects that its year-end debt at the end of 2010 will be $13.2 billion; the challenges faced by USPS in cutting costs, restructuring its workforce and networks, and range of mail products are such that the U.S. Government Accountability Office (2009c) added USPS finances to its "high-risk" list of government programs.[11]  It is also possible to envision electronic communication becoming so ubiquitous that physical mail may come to be perceived as a nuisance or inconvenience. By these points, we do not intend to imply the demise of the mail as the means for the census, but we do suggest that alternative contact strategies to achieve census participation should be a part of census research in the coming decade.

Taken together, these and other dynamics suggest that this is an opportune time to reconsider how the census is taken—to consider an approach to the 2020 census that is more than an incremental change from the general parameters of census-taking that have been used since 1970. The goal should be an efficient and effective census—one that both reduces costs and maintains (or improves) quality—that challenges long-held assumptions and draws ideas from all quarters. To determine what new census designs can

---

[11]In this designation, USPS joins the 2010 census itself, which was added to the high-risk list in March 2008; see http://www.gao.gov/press/highrisk_pressrelease_3_2008.pdf.

best respond to these dynamics, the Census Bureau needs to have a research program on census methodology that can assess which of several possible approaches to the 2020 census is most likely to provide a census responsive to these demands for reduced costs and for either equal or greater accuracy. In the remaining chapters of this report, we sketch some features of such a research program. We also suggest that commitment to major change should be matched by a bold goal, reflecting the key drivers of cost and quality. It has almost become rote to include "containing cost" as a goal of the decennial census. As a general vision for 2020, we suggest something more ambitious:

> *Recommendation 2.1:* **The Census Bureau should approach the design of the 2020 census with the clear and publicly announced goal of reducing the inflation-adjusted per housing unit cost to that of the 2000 census (subtracting the cost of the 2000 census long-form sample), while holding coverage errors (appropriately defined) to approximately 2000 levels.**

(By "appropriately defined" coverage errors, we mean quality targets for subnational net error, including by major demographic groups, as well as overall national net error.)

To be clear, we do not make this recommendation to suggest cutting census costs simply for the sake of cost-cutting. The objective is a more efficient census that, for example, uses lower-cost options like the Internet and administrative records to supplement or replace expensive follow-up operations without compromising census quality; effective research and improved cost modeling are critical to determining to what extent these efficiencies can be realized. The reason for setting a cost goal that is stark, ambitious, and public is to convey commitment to a focused 2020 census planning process.

We note that in developing cost estimates throughout the decade, the Census Bureau should be consistent in presenting its estimates. Because the decennial census no longer includes a long-form sample, the 2020 census cost goal should exclude the costs of the American Community Survey, which will increasingly take on a life of its own as a source of intercensal data on a wide range of population characteristics.

Merely slowing or curbing the rate of cost growth of the decennial census is a good goal, but we think it more advisable to target meaningful reductions in per-household cost—through leveraging new technology and methodology—without impairing quality.

**Table 2-4** Coverage Improvement Programs and Procedures, 1970–2010 Censuses—Address List Development

| Census | Original Sources of List(s) | Field Checks of List | Local Review of List |
|---|---|---|---|
| 1970 | • Tape Address Register (TAR) Areas (33% of population; city-style addresses used for mailout-mailback; urban cores of largest metropolitan areas): computerized commercial list, though about 7.2 million addresses found in "prelisting" canvass operation (see below) were added to mailout-mailback<br>• Prelist Areas (17% of population; mailout-mailback non-city-style areas [e.g., suburban rings of largest metro areas]): census enumerators developed hand-written address registers in precensus canvass operation<br>• Conventional (37% of population; other urban, suburban, and rural areas): Post Office delivered unaddressed questionnaires before enumerator visit<br>• Special Places/Group Quarters (3% of population): no advance list developed | • TAR:<br>　– Advance Post Office Check (1969, 75% of list)<br>　– Precanvass (early 1970, census enumerators checked list in hard-to-count areas of 17 metro areas)<br>　– Casing Check (post office, month before Census Day, entire list)<br>　– Time of Delivery Check (post office, entire list)<br>• Prelist:<br>　– Casing Check<br>　– Time of Delivery Check<br>• Conventional: Postenumeration Post Office Check (1970; limited to states in South; see Table 2-7) | None |
| 1980 | • TAR Areas (51% of population; mailout-mailback city-style): commercial address lists purchased between October 1978 and April 1979 | • TAR:<br>　– Advance Post Office Check (1979, entire list)<br>　– Precanvass (1980, entire list)<br>　– Casing Check (post office, 3 weeks prior to Census Day, entire list)<br>　– Time of Delivery Check (post office, entire list) | None |

| | | | |
|---|---|---|---|
| | • Prelist Areas (41% of population): census enumerators listed addresses between February and November 1979 (with instructions to canvass in a clockwise direction around every block and knock on every door), subsequently keyed into electronic form<br>• Conventional or List/Enumerate Areas (5% of population): no advance list (enumerators developed list at the same time as they obtained responses to questions)<br>• Special Places/Group Quarters (3% of population): compiled from variety of sources | • Prelist:<br>– Casing Check<br>– Time of Delivery Check | None |
| 1990 | • Address Control File (ACF) Areas (56% of population): two commercial lists<br>• 1988 Prelist Areas (26% of population): census enumerators built list in 1988<br>• 1989 Prelist or Update/Leave/Mailback Areas (10% of population): census enumerators built list in 1989<br>• List/Enumerate Areas (5% of population): no advance list<br>• Special Places List (3% of population): compiled from variety of sources | • ACF:<br>– Advance Post Office Check (APOC; 1988, entire list)<br>– Precanvass (1989, entire list)<br>– Casing Check (post office, 3 weeks before Census Day, entire list)<br>• 1988 Prelist:<br>– Advance Post Office Check (1988, entire list)<br>– APOC Reconciliation (census enumerators checked undeliverable and missing addresses in the field)<br>– Casing Check<br>• 1989 Prelist: census enumerators updated at time of delivery | • ACF: local governments given opportunity to review ACF-based housing counts by block; areas with significant discrepancies recanvassed by census enumerators (fall 1989)<br>• 1988 Prelist: same as ACF Areas |

**Table 2-4** (continued)

| Census | Original Sources of List(s) | Field Checks of List | Local Review of List |
|---|---|---|---|
| 2000 | • Master Address File (MAF) City-Style Areas (81% of population): started with 1990 ACF, updated every 6 months with Postal Service Delivery Sequence File (DSF)<br>• MAF Non-City-Style or Update/Leave/Mailback Areas (16% of population): census enumerators built list in 1998–1999<br>• List/Enumerate Areas (1% of population): no advance list<br>• Special Places List (3% of population): compiled from variety of sources | • MAF City-Style:<br>  – Block Canvass (1999, census enumerators rechecked entire list)<br>  – Reconcile (verify) LUCA changes (see next column)<br>  – Incorporate updates from Postal Service's final intensive check of DSF prior to delivery<br>• MAF Update/Leave: census enumerators updated at time of delivery | • MAF City-Style:<br>  – Local Update of Census Addresses (LUCA) Program 1998: local review of MAF; local review of Census Bureau feedback; review of local appeals by Office of Management and Budget (OMB)<br>  – New Construction LUCA (early 2000)<br>• MAF Update/Leave: LUCA Program 1999—similar to MAF City-Style Areas but later<br>• Special Places (Group Quarters): LUCA review opportunity in early 2000 |
| 2010 | • MAF City-Style Areas (76% of population): started with 2000 MAF, updated every 6 months with Postal Service DSF<br>• MAF Non-City-Style or Update/Leave/Mailback Areas (20% of population): started with 2000 MAF, updated every 6 months with Postal Service DSF<br>• List/Enumerate Areas (1% of population): no advance list<br>• Group Quarters List (3% of population): compiled from variety of sources | • MAF City-Style Areas: Address Canvass (2009, entire list); included group quarters<br>• MAF Update/Leave Areas:<br>  – Address Canvass (2009, entire list)<br>  – Census enumerators will update at time of delivery<br>• Group Quarters: list validation, Fall 2009 | • MAF City-Style:<br>  – Local Update of Census Addresses Program: local review of MAF; local review of Census Bureau feedback; review of local appeals by OMB<br>  – New Construction LUCA (early 2010)<br>• MAF Update/Leave: Local Update of Census Addresses Program, similar to MAF City-Style Areas<br>• Special Places (Group Quarters): LUCA review opportunity combined with household LUCA |

SOURCES: Anderson (2000); National Research Council (1995:App. B); U.S. Census Bureau (1976, 1989, 1993, 1995a,b, 1996, 1999).

**Table 2-5** Coverage Improvement Programs and Procedures, 1970–2010 Censuses—Publicity/Outreach

| Census | Advertising | Pre-Census Day Outreach | Post-Census Day Outreach |
|---|---|---|---|
| 1970 | • Print media public service advertising, January–April (Advertising Council selected vendor, worked pro bono): publicity releases in 23 languages and "Were You Counted?" forms in 8 languages other than English; billboards in 20 metropolitan areas and placards distributed to transit companies by Advertising Council; posters and flyers distributed to school systems and post offices<br>• Radio and television advertising coordinated by Advertising Council; public service announcements by entertainment personalities; voice-over public service announcements at end of programs on two television networks; recorded spots distributed to radio stations, including Spanish versions<br>• Bureau-appointed "public information specialists" in New York, Philadelphia, Chicago, Detroit, and Los Angeles; Commerce Department publicity personnel in four other cities | • Encouraged communities to set up "Complete Count Committees" | • Assistance Centers to call or visit in 20 cities<br>• Missed Persons Campaign ("Please Make Sure I Am Counted in the Census" cards—in English, Spanish, or Chinese—requesting minimal information distributed to carryouts, barbershops, etc., and matched to census to verify)<br>• "Were You Counted?" Campaign (short-form questionnaire printed in numerous newspapers for readers to clip and return if they or household members believed they had been missed; forms also distributed to local governments for distribution) |

**Table 2-5** (continued)

| Census | Advertising | Pre-Census Day Outreach | Post-Census Day Outreach |
|---|---|---|---|
| 1980 | • Public service advertising (Advertising Council selected Ogilvy & Mather as vendor, working pro bono); idea of seeking $40 million appropriation for paid advertising considered but ultimately rejected<br>  – Subcampaigns: Pre-Census Day ("Answer the Census—We're Counting on You"), post-Census Day ("It's Not Too Late"), business manager focus (promoting employee response), and enumerator recruitment<br>  – Print media campaign included outdoor posters and transit cards, including posters in about a dozen languages; Census Bureau publicity office contacted 75 top cartoonists to encourage them to use census themes in editorial cartoons and comic strips<br>  – Radio and television campaign included Spanish-language spots and about 44 celebrity public service announcements<br>• Broadcasters Census Committee of '80 (station owners and managers) convened by Census Bureau and Department of Commerce to promote prime placement of messages<br>• Special outreach effort to minority media organizations | • Encouraged communities to set up about 4,000 complete count committees<br>• Request to "key persons" in 300 census statistical areas—typically city or county planning officials—to volunteer time as liaison between the Census Bureau and local committees providing input on small-area geographic designations (tracts, neighborhoods, etc.)<br>• Outreach to about 600 major national service organizations and labor groups beginning in July 1979<br>• Census Bureau publicity office contacted trade associations to encourage them to communicate census messages, including American Society of Association Executives and Food Marketing Institute<br>• Partnership with selected major businesses, including five largest grocery chains, General Cinema Corporation, and Goodyear<br>• Outreach to members of Congress to convey census messages and to encourage them to tape radio and television appeals for census participation<br>• School Project distributed census-related materials for use in grades 4–12 | • Specific "It's Not Too Late" theme in public advertising campaign<br>• "Were You Counted?" campaign included advertisements in local newspapers (available in 33 languages), short-form questionnaire for printing in local newspapers (for readers to clip and return), and distribution of special forms to local governments and complete count committees |

| 1990 | | | |
|---|---|---|---|
| | • Public service advertising (Advertising Council selected Ogilvy & Mather as main vendor, working pro bono, as in 1980)<br><br>• Four minority advertising firms recruited by Advertising Council to produce public service announcements and advertisements focused on black, Hispanic, Asian and Pacific Islander, and Puerto Rican communities<br><br>• Broadcasters Census Committee of '90 established by National Association of Broadcasters to encourage support and broadcast of census messages in electronic media; differed from 1980 version by including radio managers as well as television<br><br>• Efforts to secure pro bono ad slots and segments on major cable networks as well as broadcast networks<br><br>• Publicized "kickoff" events for publicity campaign in various ethnic communities in February and March 1990 | • Census Awareness and Products Program developed partnerships with local community organizations<br><br>• Community awareness specialists assigned to all regional census centers<br><br>• Concerted effort on "census as news"—Census Bureau publicity ofice active in distributing press releases and feature articles; effort included intense publicity on S-Night operations<br><br>• Mayors' cooperation program: publicized visits by high-level Bureau staff to 35 mayors<br><br>• Government Promotion Handbook sent to all local governments to encourage formation of complete count committees<br><br>• Census Education Project on larger scale than 1980; teaching kit sent to all elementary and secondary schools, aimed at grades 4–12; Bureau also supported teacher training workshops<br><br>• Public Housing Initiative: working with local public housing authorities, residents of public housing complexes were sought for outreach and to work as urban update/leave enumerators<br><br>• Information booklet mailed to almost 400,000 churches<br><br>• Work with National Head Start Initiative to promote message placement and encourage Head Start families and staff to seek work as census takers | • "Were You Counted?" campaign (questionnaire printed in various periodicals for readers to clip and return; electronic media messages emphasized ability to call telephone assistance line)<br><br>• "Thank You America" program: certificates of appreciation and plaques distributed to active individuals and organizations; direct "thank you" message also included in mailing of final population counts to local governments |

**Table 2-5** (continued)

| Census | Advertising | Pre-Census Day Outreach | Post-Census Day Outreach |
|---|---|---|---|
| 2000 | • Paid advertising, November–June (Young and Rubicam, Inc., contractor), 17 languages<br>• "Census 2000 Road Tour" included about 2,000 stops at community events | • Partnership and Marketing Program (more extensive)<br>  – About 650 partnership specialists hired to maintain partnerships with local governments and community organizations<br>  – Supported creation of complete count committees<br>• Census in the Schools (K-12)<br>• Kits distributed to Head Start and adult citizenship and literacy training centers<br>• Direct mail campaign included multiple cues and reminders (e.g., advance letter, reminder/"thank you" postcards) | • Be Counted Campaign (media announcements encouraged people to send in a special form, placed in public locations) |
| 2010 | • Paid advertising (DraftFCB as prime contractor, with several partner agencies to reach specific markets), 28 languages (originally 14)<br>• Integrated Communications Campaign Plan; includes advertising in new and nontraditional media (e.g., Internet, grocery stores)<br>• Extensive audience segmentation research done in planning campaign, including analysis of American Community Survey data<br>• "Census Road Tour" | • Partnership Program (even more extensive)<br>• Census in the Schools (K-12) (originally K-8)<br>• Bilingual English-Spanish questionnaire mailed to residents in selected areas | • Replacement questionnaires mailed to selected households<br>• Be Counted Campaign (media announcements will encourage people to send in a special form) |

SOURCES: See Table 2-4.

**Table 2-6** Coverage Improvement Programs and Procedures, 1970–2010 Censuses—Initial Enumeration Methods

| Census | Type of Enumeration | Questionnaire/Mailing Package | Office/Processing Structure |
|---|---|---|---|
| 1970 | • Mailout-mailback (postal service delivered addressed questionnaires; respondents mailed them back)<br>• Conventional (postal service delivered unaddressed questionnaires; census enumerators picked up and completed)<br>• Special Places (Group Quarters) visited and enumerated<br>• Transients (in hotels, campgrounds, etc.) enumerated on "T-night" | • Computer-readable questionnaire | • 12 permanent regional offices (Atlanta, Boston, Charlotte, Chicago, Dallas, Denver, Detroit, Los Angeles, New York, Philadelphia, St. Paul, Seattle)<br>• 1 temporary area office (San Francisco)<br>• 393 temporary local offices<br>• Suitland headquarters<br>• Processing centers in Jeffersonville, IN, and Pittsburg, KS |
| 1980 | • Mailout-mailback (as in 1970)<br>• List/enumerate (census enumerators developed list and enumerated residents)<br>• Special Places (Group Quarters) visited and enumerated<br>• "T-night" operation on March 31 to count transients or long-term residents of hotels, motels, and tourist homes<br>• Overnight "M-night" ("mission night") operation on April 8–9 to count people at places like missions and flophouses, short-term detainees in jail, and people spending night in transportation terminals | • Computer-readable questionnaire | • 12 permanent regional offices (same as 1970 and continuing to present, replacing St. Paul with Kansas City, KS)<br>• 12 regional census centers, located in regional office cities<br>• 412 local offices<br>• Suitland headquarters<br>• Processing centers in Jeffersonville, IN; Laguna Niguel, CA; and New Orleans, LA |

**Table 2-6** (continued)

| Census | Type of Enumeration | Questionnaire/Mailing Package | Office/Processing Structure |
|---|---|---|---|
| 1980 (cont.) | • Casual Count: in inner-city areas, census enumerators visited employment or welfare offices during daylight hours and major gathering places (pool halls, street corners, bars, etc.) in early evening; locations chosen based on suggestions from local officials, and operation generally took place over 1–2 weeks<br><br>• "Overseas Travel Reports" distributed to international air and ship lines for distribution between March 15 and April 1 (for mailback by respondent) to try to count U.S. residents in transit | | |
| 1990 | • Mailout-mailback (as in 1970)<br>• Update List/Leave (census enumerators delivered addressed questionnaires and updated list; respondents mailed back)<br>• List/Enumerate (census enumerators developed list and enumerated residents)<br>• Urban Update/Leave and Urban Update/Enumerate operations applied rather than mailout-mailback in urban blocks containing predominantly public housing developments or boarded-up buildings, respectively | • Computer-readable questionnaire<br>• Reminder postcard 1 week after questionnaire delivery (in mailout-mailback and Update List/Leave areas) | • 12 permanent regional offices<br>• 12 regional census centers (12 located in regional office cities, plus temporary center in San Francisco)<br>• 449 local offices<br>• Suitland headquarters<br>• Backup computer center in Charlotte, NC<br>• Jeffersonville, IN, National Processing Center<br>• 7 processing offices for data capture (Albany, NY; Austin, TX; Baltimore, MD; Jacksonville, FL; Jeffersonville, IN; Kansas City, MO; San Diego, CA) |

- Special Places (Group Quarters) visited and enumerated
- Transients enumerated on "T-night," March 31, focusing on transient places such as campgrounds and marinas
- Shelter and Street Night ("S-night"): replacement and expansion of "M-night" and casual count operations of 1980; two phases conducted overnight between March 20 and 21, first counting shelter inhabitants and then a "street" count of people in open public locations preidentified by local governments

- 12 permanent regional offices
- 12 regional census centers
- 520 local offices
- Suitland headquarters
- National Processing Center (Jeffersonville, IN)
- 3 contractor-operated data capture centers

**2000**

- Mailout-mailback (as in 1970)
- Mailout-mailback Conversion to Update List/Leave and Urban Update List/Leave (originally designated for mailout, census enumerators delivered addressed questionnaires and updated list; respondents mailed them back)
- Update List/Leave (as in 1990)
- Rural Update List/Enumerate (originally designated for Update List/Leave, census enumerators delivered addressed questionnaires, updated list, and enumerated respondents)
- List/Enumerate (census enumerators developed list and enumerated residents; earlier in remote Alaska)
- Special Places (Group Quarters) visited and enumerated
- Transients enumerated on "T-night"
- Service-Based Enumeration (SBE) (people enumerated at selected locations, such as shelters)

- Advance letter (in mailout-mailback areas)
- User-friendly optical-scanner-readable questionnaire
- Reminder postcard 2 weeks after questionnaire delivery (in mailout-mailback and Update List/Leave areas)
- Internet response option available in place of mailback
- Telephone response option available in place of mailback (also to answer questions)

**Table 2-6** (continued)

| Census | Type of Enumeration | Questionnaire/Mailing Package | Office/Processing Structure |
|---|---|---|---|
| 2010 | • Mailout-mailback (as in 1970)<br>• Urban Update List/Leave (designed for urban areas with lots of P.O. boxes, apartments with central mail drop, etc.)<br>• Update List/Leave (as in 1990)<br>• Update List/Enumerate (designed for colonias, areas with large numbers of vacant units, and American Indian tribal areas)<br>• Remote List/Enumerate (census enumerators will develop list and enumerate residents; to be conducted earlier in remote Alaska)<br>• Special Places (Group Quarters) will be visited and enumerated; in addition, there will be an advance visit<br>• Transients will be enumerated on "T-night"<br>• Service-Based Enumeration (SBE) (people enumerated at selected locations, such as shelters) | • Advance letter (in mailout-mailback areas)<br>• User-friendly optical-scanner-readable questionnaire<br>• Bilingual Spanish-English questionnaire (to be sent to 13 million addresses in areas with high concentrations of Hispanics)<br>• Reminder postcard 2 weeks after questionnaire delivery (in mailout-mailback and Update List/Leave areas)<br>• Second questionnaire to be sent to everyone in hard-to-count areas and to mail nonrespondents elsewhere<br>• Telephone response option available in place of mailback (also to answer questions) | • 12 permanent regional offices<br>• 12 regional census centers<br>• 520 local offices<br>• Suitland headquarters<br>• National Processing Center (Jeffersonville, IN)<br>• 2 contractor-operated data capture centers |

SOURCES: See Table 2-4.

**Table 2-7** Coverage Improvement Programs and Procedures, 1970–2010 Censuses—Follow-Up of Mail Returns

| Census | Field Follow-Up | Telephone Follow-Up |
|---|---|---|
| 1970 | • Report of Living Quarters Check (field follow-up of households in small multiunit structures reporting more units than on the census list) | |
| 1980 | • Report of Living Quarters Check<br>• Dependent Roster Check (more residents reported on front of questionnaire than on inside pages)<br>• Whole Household Usual Home Elsewhere Check—follow-up of questionnaires where question asking whether "everyone here is staying only temporarily and has a usual home elsewhere" was answered in affirmative and additional home address(es) listed on back on form<br>• Personal visit follow-up of cases where telephone follow-up failed | • Follow-up for missing information (item nonresponse), households apparently returning duplicate questionnaires, or cases where field "Followup 1" (see Table 2-8) efforts failed |
| 1990 | • Whole Household Usual Home Elsewhere Check<br>• Follow-up for missing information | • Follow-up for missing information (as in 1980) |
| 2000 | | • Coverage Edit Follow-Up (returns that reported more or fewer people in Question 1 than on the individual pages; returns with 7 or more household members [the form only included room for 6]; and returns with Question 1 blank and exactly 6 people with individual information provided) |
| 2010 | | • Coverage Follow-Up (same as 2000, with the possible addition of duplicates discovered in nationwide matching of the census against itself, cases in administrative records that do not match a return, and responses to two new coverage questions that indicate possible problems) |

SOURCES: See Table 2-4.

**Table 2-8** Coverage Improvement Programs and Procedures, 1970–2010 Censuses—NRFU (Nonresponse Follow-Up) and Post-NRFU

| Census | Program |
|---|---|
| 1970 | • NRFU (mailback areas)—2 visits before going to Last Resort (neighbor, landlord, etc.)<br>• National Vacancy Check—Sample survey of 13,500 housing units originally classified as vacant; results used to reclassify 8.5% of all vacant units as occupied and impute people into them<br>• Post-Enumeration Post Office Check (PEPOC)—Postal Service checked enumerator address lists in conventionally enumerated areas of 16 Southern states; census enumerators followed up sample of missed addresses; results used to impute housing units and people into the census based on postal review of addresses listed by enumerators in conventional enumeration areas |
| 1980 | • NRFU (mailback areas)—3 visits before going to Last Resort; conducted in two waves:<br>  – "Followup 1" (April 16–planned end on May 13): Obtain questionnaires from households not returning forms by 2 weeks after Census Day and to check on vacant units<br>  – "Followup 2" (May 22–July 7): Begin with Vacant/Delete Check—100% reenumeration of units originally classified as vacant or nonresidential—and move on to follow-up still-missing questionnaires, households where interview attempts in "Followup 1" failed, and persons found using nonhousehold sources (see below). Followup 2 also included recanvass of enumeration districts in 137 local offices where local review suggested problems with prelist addresses (see Table 2-4)<br>• Post-Enumeration Post Office Check (PEPOC)—100% recheck of address list and enumeration of missed units in conventionally enumerated areas<br>• Prelist Recanvass—Address list rechecked in Prelist areas (in some areas only selected enumeration districts rechecked)<br>• Local Review—Local officials provided with preliminary housing unit and population counts to indicate problem areas for rechecking<br>• Nonhousehold-Sources Program—Administrative lists (driver's license records, immigration records, and New York City public assistance records) matched to census records for selected census tracts in urban district offices and missing people followed up |
| 1990 | • NRFU (mailback areas)—6 visits (3 in-person, 3 telephone) before going to Last Resort<br>• Vacant/Delete Check—100% reenumeration of units originally classified as vacant or nonresidential (except for seasonal vacants and units classified by 3 or more operations as vacant)<br>• Housing Coverage Check (recanvass)—Reenumeration of over 500,000 blocks with 15 million housing units identified as problematic from various sources (e.g., comparisons with building permit data)<br>• Local Review—Local officials provided with preliminary housing unit and population counts to indicate problem areas for rechecking (about 150,000 blocks were recanvassed) |

| | |
|---|---|
| 1990 (cont.) | • Parolee/Probationer Check—Questionnaires were distributed to parole and probation officers to distribute to those under their jurisdiction |
| | • Reenumeration of households with only one member in 24 local offices and partial reenumeration of households in 7 local offices in NJ—Carried out because of allegations of fraud by enumerators |
| 2000 | • NRFU (mailback areas)—6 visits before going to Last Resort (as in 1990) |
| | • Vacant/Delete Check—100% reenumeration of units originally classified as vacant or nonresidential (except for seasonal vacant units and units classified by 2 or more operations as vacant) |
| | • Coverage Improvement Follow-Up (CIFU)—Revisit of addresses added in Update/Leave for which no questionnaire was mailed back; New Construction LUCA Program addresses; blank mail returns; late Postal Service Delivery Sequence File address additions |
| | • Field Verification—Field check of various addresses that were not resolved after CIFU |
| | • Ad Hoc Master Address File (MAF) Unduplication—Comparisons with building permits, etc. suggested substantial duplication of addresses on the MAF (even after unduplication operation prior to NRFU); matching program and subsequent examination identified 1.4 million housing units to delete |
| | • Reenumeration of 1 local office and partial reenumeration of 7 offices—Carried out because of evidence of incompetence or fraud |
| 2010 | • NRFU (mailback areas, new construction addresses)—6 visits before going to Last Resort (as in 1990) |
| | • Vacant/Delete Check—100% reenumeration of units originally classified as vacant or nonresidential (as in 2000) |
| | • Coverage Follow-Up (CFU)—Telephone-only follow-up, which may include NRFU as well as mail returns—see Table 2-7 |
| | • Field Verification—Field check of various addresses that are not resolved after CFU |
| | • Last minute, ad hoc checks likely based on past experience |

SOURCES: See Table 2-4.

# – 3 –

# Census Bureau Research, Past and Present

H AVING CONCLUDED IN CHAPTER 2 that serious attention to cost and quality must drive the planning for the 2020 census, we describe our recommendations in the following two chapters. In this chapter, we critique the Census Bureau's existing program for research—exemplified by the 2010 Census Program of Experiments and Evaluations (CPEX)—both by comparison with the Bureau's past efforts and through articulation of the gaps in its current strategies toward research. This chapter then provides general guidance for rethinking census research (and, with it, the approach to the 2020 census); Chapter 4 turns to the practical issues of structuring and scheduling research in advance of the decennial census.

Section 3–A presents our critique of current and recent trends by the Census Bureau in its research programs; it builds from the context provided by two detailed appendices at the end of this report. Appendix A describes the precensus testing programs and formal research (experimentation and evaluation) programs of the 1950–2000 censuses. It also describes the testing earlier in this decade related to the 2010 census. Appendix B then describes the 2010 CPEX program in detail. Section 3–B rounds out this chapter by laying out what we think are key steps in improving the substance of Census Bureau operational research.

It is important to note two caveats about the historical review of research in Appendix A (and, indeed, throughout this report). First, in summarizing research activities in the appendix, our concern is more in inventorying the types and varieties of activities that have characterized Census Bureau research and less in completely documenting their results. This is partly due to availability of information—particularly for the earlier censuses, methodological details are scant in the literature—and partly because documenting the full "results" of major field tests such as census dress rehearsals is simply beyond the scope of our project. Our interest in the appendix and this chapter is in describing general contours and features of census research and not on assessing the merits (or lack thereof) of each specific activity.

Second—and more fundamentally—we deliberately do not delve into the details of the coverage measurement programs that have accompanied the decennial censuses (save for formative developments in the earliest decades). We concur with our predecessor National Research Council panels that have found the Census Bureau's coverage measurement programs to be generally of high quality. In particular, the Panel to Review the 2000 Census concluded that coverage measurement work in 2000 "exhibited an outstanding level of creativity and productivity devoted to a very complex problem" and that it showed "praiseworthy thoroughness of documentation and explanation for every step of the effort" (National Research Council, 2004a:244, 245). We also direct interested readers to the final report of another National Research Council (2008a) panel that had coverage measurement research as its explicit charge, which is not the case for our panel. Given our charge, we have focused our own analysis principally on research and development related to census operations; accordingly, it is important to note that our comments on "census research" in what follows should not be interpreted as applying to the Census Bureau's extensive body of coverage measurement research.

## 3–A  CURRENT RESEARCH: UNFOCUSED AND INEFFECTIVE

In our assessment, the Census Bureau's current approach to research and development (R&D), as it applies to decennial census operations, is unfocused and ineffective. The Census Bureau's most recent research programs suffer from what one of our predecessor panels described as a "serious disconnect between research and operations in the census processes" (National Research Council, 2004b:45):

> Put another way, the Census Bureau's planning and research entities operate too often at either a very high level of focus (e.g., articulation of the "three-legged stool" concept for the 2010 census) or at a microlevel that tends toward detailed accounting without much analysis. . . . What is lacking is research, evaluation, and planning that bridges these two levels, synthesizing the detailed results in order to determine their implications for planning while structuring high-level operations in order to facilitate meaningful detailed analysis. Justifying and sustaining the 2010 census plan requires both research that is forward-looking and

strongly tied to planning objectives, and rigorous evaluation that plays a central role in operations rather than being relegated to a peripheral, post hoc role.

In our assessment, the 2010 CPEX exemplifies these problems, although the problem is broader than the specific experiments, evaluations, and assessments outlined in Appendix B. As the quotation from the predecessor panel suggests, our critique applies to Census Bureau research writ larger, including the census tests conducted between 2000 and 2010 and the evaluations and experiments of the 2000 census.

### 3–A.1  Legacy of Research

In Appendix A, we outline the precensus testing activities and formal research, experimentation, and evaluation programs of the 1950–2000 censuses. Together, they provide a picture of the Census Bureau's major research programs on the decennial census, from the buildup to the 1950 census through the 2008 dress rehearsal for the 2010 census. We refer to specific points in these narratives throughout this chapter, and also note some general impressions from the flow of census research over the years. The first such observation from past census research is that the lack of focus evident to us in the Bureau's current strategy was not always the case. Indeed, the Census Bureau has in the past been a place where major technological improvements and major data collection changes have been successfully executed through careful (but innovative) research and pathbreaking theoretical work.

Arguably the best example of R&D driving change in the process of census-taking—a string of related research projects building toward a major operational goal—is the switch from enumerator-conducted personal visits to mailed, self-response questionnaires as the primary mode of data collection. Now that mailout-mailback methodology has become so ingrained, it can be difficult to fully grasp how seismic a shift in methodology the change to mail was for the census. However, the magnitude of the shift can be inferred from chronicling the careful, deliberate program of testing and experimentation that preceded the change. As the Census Bureau's procedural history of the 1970 census (U.S. Census Bureau, 1976) notes, mail was used for some Census Bureau activities as early as 1890—predating the establishment of the Bureau as a permanent office. In 1890, questionnaires concerning residential finance were mailed to households with a request for mail return; the same was repeated in 1920, and a similar mail-based program of income and finance questions was also used in 1950. Supplemental information on the blind and the deaf was requested by mail in the 1910, 1920, and 1930 censuses, and a mail-based "Absent Family Schedule" was used for some follow-up work in 1910, 1930, and 1940. The direct path to mail as the primary mode for census collection probably begins with the "Advance Schedule of Population" that was delivered to households in 1910; this form was meant to acquaint households with the topics of the census but was not meant to be completed by the householders (a similar advance form was used in the agriculture census conducted in 1910). Following World War II, and often in conjunction with special censuses requested by cities and towns, the Census Bureau initiated a set of experiments and tests of mailout or mailback methods; one such test was conducted in 1948 (Little Rock, AR; Section A–1.a), another as a formal experiment of the 1950 census in which households in Columbus, OH, and Lansing, MI, had questionnaires distributed to them by enumerators prior to the 1950 census with instructions to complete and mail them on Census Day (U.S. Census Bureau, 1966:292; see also U.S. Census Bureau, 1955:5). Similar tests were performed in 1957, 1958, and 1959, with the January 1958 test in Memphis, TN, adding field follow-up of mailed census returns as a check on quality (Section A–2.a).

In the 1960 census, households were mailed an "Advance Census Report," which they were asked to fill out but *not* return by mail. Instead, enumerators visited the household to collect the forms and transcribe the information onto forms more conducive to the optical film reader then used to process census data. If a household had not completed the advance form, the residents were interviewed directly by the enumerator. Based on the successful use of the mailout questionnaire in 1960, Congress enacted a brief but powerful amendment to census law in 1964: P.L. 88-530 struck the requirement that decennial census enumerators must personally visit every census household. Even though mail methods had been tested and the required legal authorization had been obtained, mailout-mailback methods were subjected to further testing prior to the 1970 census, as Section A–3.a describes. These designed tests of mail procedures escalated in size and complexity from a relatively small community (Fort Smith, AR) to a large central city (Louisville, KY), to known hard-to-count areas (parts of Cleveland, OH, that experienced enumeration problems in 1960 and ethnic communities in Minnesota and New York). In the 1970 census, questionnaires were mailed to about 60 percent of all housing units, focusing on major urbanized areas; a formal experiment conducted during the 1970 census (Section A–3.b) expanded mailout-mailback methods to more rural areas in 10 local offices, anticipating wider use of mail methods. The percentage of the population in the mailout-mailback universe has grown in subsequent censuses to include 81 percent of the population in 2000, with another 16 percent receiving questionnaires from census enumerators to be mailed back (see below).

Other notable examples in which R&D (in the form of census tests, experiments, and evaluations) drove important developments in the census process include:

- *Refinement of residence rules for college students:* Prior to the 1950 census, the results of test questions in Current Population Survey (CPS) supplements and special censuses of cities contributed to the Bureau's reversing its rule on counting college students. As described by the National Research Council (2006:Sec. 3–B.1), census practice since 1880 had favored counting college students at their parental homes. The test results prior to 1950 contributed to the conclusion that college students enrolled at schools away from home were frequently omitted in parental household listings, so the Census Bureau reverted to the 1850 census approach of counting college students at their school location.

- *Development of enumeration strategies for nonmail areas:* In the 2000 census, blocks and other geographic regions were designated to be covered by one of nine types of enumeration areas (TEAs)—essentially, the method for making initial contact with census respondents—with mailout-mailback being the most common TEA. *Update-leave* enumeration for areas without city-style addresses, in which enumerators checked and updated address list entries during their visits but simply left a census questionnaire for respondents to return by mail, was first tested in a significant way in one of the experiments of the 1980 census; five matched pairs of district offices were selected for comparison using this technique (Section A–4.b). The March 1988 dress rehearsal in St. Louis, MO, added a variant of the strategy—*urban update/leave*, targeting hard-to-enumerate areas for personal enumerator visit—that was added to the 1990 census, evaluated, and judged to be a valuable technique (Sections A–5.a and A–5.b). The Bureau's response to an unforeseen problem in the same 1988 dress rehearsal also led to enduring changes in practice. Nine counties in the east central Missouri test site were initially thought to be amenable to mailout-mailback but a pre-census check suggested high levels of undeliverable addresses. Hence, the Bureau swapped strategies for such areas; the same flexibility in approach applied to the 2000 census, in which *mailout-mailback conversion to update-leave* was one of the TEAs.

### 3–A.2   Flaws in Current Census Research and the 2010 CPEX

In this section, we briefly describe the principal deficiencies that we observe in the Census Bureau's current approach to research and in the 2010 CPEX in particular. In our assessment, shortcomings in the Census Bureau's research strategy need to be overcome immediately in order to foster an effective research program for the 2020 census.

However, at the outset of this discussion, it is important to note—and commend—revisions to 2010 census research that were announced as this report was in the late stages of preparation. In congressional testimony in October 2009, Census Bureau Director Robert Groves (2009:9) described three research programs that he initiated following his own evaluation of 2010 census preparations:

> We will develop and implement a Master Trace Project to follow cases throughout the decennial census cycle from address listing through tabulation so that we have a better research base for planning the 2020 Census. We also will be conducting an Internet measurement re-interview study, focused on how differently people answer questions on a web instrument from a paper questionnaire. Finally, we will mount a post-hoc administrative records census, using administrative records available to the Census Bureau. All of this will better position us for the developmental work we must conduct to improve future decennial census operations.

In committing to retain 2010 census operational data and to more aggressively evaluate the quality of administrative records data relative to census returns, the director's proposed programs are responsive to recommendations made in our letter report (Part III of this volume). The proposed Internet reinterview study stops short of testing Internet response in the census, but does at least put the Bureau in the position of testing Internet response to a census instrument. We commend these developments and look forward to their completion. We also note that they are also partially responsive to the recommendations and guidance in the balance of this chapter. That said, important problems remain in the Bureau's general approach to research, as we now describe.

### Lack of Relevance to Cost and Quality Issues

Although effects on cost and quality were listed by the Bureau as primary criteria for choosing studies for the 2010 CPEX, the final slate of experiments and evaluations described in Appendix B and analyzed in our letter report (Part III of this volume) seem ill-suited to inform choices that would meaningfully affect either cost or quality. Of the experiments:

- Only the nonresponse follow-up (NRFU) Contact Strategy Experiment appears clearly motivated by an attempt to reduce the cost of a high-cost operation without impairing census quality. However, even that experiment promises to stop short of providing comprehensive information on the cost-benefit trade-offs of suspending follow-up contacts after some number (more or less than 4–6) of attempts. Because enumerators will know in advance how many attempts are possible for a household, the experiment presents the opportunity for enumerators to game the system: to try particularly hard in early approaches at a 4-contact household or be slightly more casual in early attempts at a 6-contact household. The experiment may provide insight into how enumerators deal with preset rules and conditions but not a true measure of NRFU yields at each possible contact opportunity.

- The Deadline Messaging/Compressed Schedule Experiment may rightly be said to have some bearing on census costs, to the extent that it helps provide mailing package cues and prompts that may boost the mail return rate (and thus reduce the more costly NRFU workload). The compressed schedule portion of the experiment could be argued to promote higher quality by pushing data collection closer to the actual census reference date of April 1, but the impact of a one-week shift on the quality of resulting data is most likely negligible.[1]

- The Alternative Questionnaire Experiment (AQE) is heavily focused on refinements to the measurement of race and Hispanic origin— important data items to be sure, but ones for which an objective truth is both unknown and unknowable, subject as it is to individual concepts of self-identity. Hence, the experiment may suggest whether different treatments yield different levels of reporting in specific categories, but it is impossible to say whether "different" is the same as "higher quality." By comparison, only a single panel in the AQE focuses on the quality of information about residence and household count—the information that represents the constitutional mandate for the census.

- As we noted in our letter report, the Confidentiality/Privacy Notification Experiment is, if anything, contrary to the goals of reducing cost and improving quality. Its paragraph treatment raises the possibility of mixing census information with data from other government agencies—i.e., the use of administrative records in census processes— in an ominous manner. Since the experiment includes only the single alternative wording, it creates a situation where respondents may react negatively but relatively little is learned about public sensitivity to records use.

### Undue Focus on "Omnibus" Testing Slots

One observation that is clear from comparing previous census research programs (e.g., Sections A–1.a–A–4.b) with the 2000 and 2010 research pro-

---

[1]A treatment group in the 2006 Short Form Mail Experiment (see Section A–7) that used a compressed mailing schedule showed similar levels of nonresponse to questionnaire items on housing tenure, age, Hispanic origin, race, and sex compared with a control group. The compressed schedule group had a statistically significant difference (decrease) in leaving the total household count blank compared with the control, and also appeared to increase reporting of new babies and reduce the tendency for respondents to omit themselves from the questionnaire (Martin, 2007), although how these effects are specifically generated by a one-week difference in questionnaire mailout is unclear.

grams is that the Bureau used to be considerably more flexible in the forms of research studies it undertook. Small, targeted tests in selected sites used to be more frequent; again, the example of the final gear-up to mailout-mailback methodology in 1970 (Section A–3.b) is instructive, with a series of tests escalating in size and scope from small communities to dense urban centers. The Census Bureau also made considerable use of special censuses commissioned by individual localities as experimental test beds; costs of the tests were thus shared by the Census Bureau and the sponsoring locality, and the locality had a tangible product—fresh population data—as an incentive for cooperating with the experimental measures. The use of special censuses for such purposes seems to have ended—perhaps understandably so—when the city of Camden, NJ, sued the Census Bureau over the results of a September 1976 test census in that city, occasioning several years of legal wrangling (Section A–4.a). In the past, the Census Bureau was also more willing to use other surveys for testing purposes—particularly the Bureau-conducted Current Population Survey, which was used to test items for the 1950 and 1960 censuses.

In the most recent rounds of research and experimentation, selected studies seem to have been chosen more based on the availability of testing "slots" (e.g., something that could be tested using only the mail as part of a single, omnibus, mail-only experiment in a year ending in 3) than on looming questions and operational interests. The recent cycle of mail-only tests in years ending in 3 or 5, tests involving a field component in years 4 and 6, and a dress rehearsal in year 8 has the advantage of keeping the various parts of census processing in fairly constant operation, so that there is no need to completely rebuild field operations from scratch. But a too-strong focus on these single-shot testing slots has led to poor design choices. For example:

- The National Research Council (2004b:227) argued that the Census Bureau's decision to fuse a test of alternative response technologies (i.e., paper, Internet, or telephone) to a mail questionnaire with numerous modules on race and Hispanic origin question wording in the 2003 National Census Test "was likely one of convenience" rather than one intended to produce meaningful results. The availability of a nationally representative sample seemed to have trumped attention to "power analysis to determine the optimal sample sizes needed to measure effects to desired precision" or "more refined targeting of predominantly minority and Hispanic neighborhoods" where the revised race questions would provide the most information.

- Early plans for the mail-only 2005 National Census Test included experimental panels of different presentations of the instructions and wording of census Question 1 (household count). However, those early plans failed to include any relevant control group—either the question as presented in the 2000 census or the modified question used in a 2004 test—making it impossible to judge effectiveness of the experimental treatments compared with a baseline. After this deficiency was identified at a meeting of the Panel on Residence Rules in the Decennial Census, the test plan was altered to include a control (National Research Council, 2006:205).

In a regime of large omnibus tests, topics that might best or more accurately be handled in a series of smaller, focused tests are forced into a larger design, without promise that the omnibus test will be able to distinguish between fine-grained alternatives. Those omnibus census tests that involve a field component also suffer from an important limitation. They are meant to be census-like to the greatest extent possible in order to utilize the complete census machinery. But an explicit proviso of the modern census tests is that no products are released, most likely to maintain consistency with the Bureau's reluctance in recent decades to use locally sponsored special censuses as experimental opportunities. But the result of this practice is a major operational test that is "census"-like save for the fact that it is not actually a census: such trials provide participating localities with no tangible product or benefit. Try though the tests do to create census-type conditions, localities have little incentive to provide unfettered support other than a sense of civic duty.

Finally, the shift in recent years to omnibus tests has created another fundamental flaw in Census Bureau research: almost of necessity, the tests can not build from each other. In previous decades, "chains" of related tests can be seen. For instance, the major census tests in Yonkers, Indianapolis, and Memphis in 1957–1958 (Section A–2.a) all involved use of a two-stage interview process, with another enumerator or supervisor rechecking results; based on experience in one of the tests, approaches in the later tests were varied. By comparison, the large-scale tests of recent years take longer to design, longer to field, and longer to analyze—and leave few resources for subsequent tests. With few exceptions, the results of recent census tests have been unable to follow directly from the experience of their predecessors simply because the results of the earlier tests had not been processed.

### Failure to Utilize Current Methods in Experimental Design

A criticism related to the increased reliance on a smaller number of large, omnibus tests is a lack of attention to some fundamentals of experimental design. In an attempt to be all-inclusive, the experimental designs of recent decennial census experiments—including those of the 2010 CPEX—do not take proper account of factors affecting the response and strength of the expected treatment effect, and, as a result, the findings from some experiments have been inconclusive. As we have already noted, the 2003 National Census Test fused together two broad topics—alternative response methodologies and variants on race and Hispanic origin questions—mainly to fill topic "slots" on the planned mailout-only test. Combining the two and distributing the sample led not only to the comparison of extremely subtle treatments in the race and Hispanic-origin segments, but also to the omission of relevant treatment groups on the response method portion (i.e., a treatment to "push" Internet use exclusively, instead of a group encouraging either Internet or telephone response).

There are several commonly used techniques that the Census Bureau does not typically employ in its experiments and tests that could provide important advantages over the methods currently used. For example, fractional factorial designs are extremely useful in simultaneously testing a number of innovations while (often) maintaining the capability of separately identifying the individual contributions to a response of interest. Such a methodology would be well suited to the problem of census innovation, since there are typically a small number of replications and a relatively large number of factors being modified. This problem is very typical of the large census tests that often need to examine a number of simultaneous changes due to limited testing opportunities.

The problem of confounding is well known, yet there are examples of experiments carried out by the Census Bureau, either during the decennial census or during large-scale census tests, in which the experiments have generated very uncertain results due to the simultaneous varying of design factors in addition to those of central interest. For example, the ad hoc Short Form Mail Experiment in 2006 (see Section A–7) took as a main objective determining whether a compressed mailing schedule and specification of a "due date" hastened response, but—by design—the experiment treated deadline and compressed scheduling as a single, combined factor and so could not provide insight as to which change was most effective.[2] The proposed Deadline Messaging/Compressed Schedule experiment in the 2010 CPEX shows similar features and flaws. The message treatments shown in Table B-1 test subtle variations of an appeal made in a short paragraph in a cover letter and reminder postcard with blunter changes made to other parts of the mailing package. Whether a quicker response is due to the appeal to save taxpayer funds by mailing the questionnaire or to the explicit "Mail by

---

[2]Describing the design, Martin (2007:12) argues that, "practically speaking, it is not feasible to implement a deadline without also moving up the mailing schedule. It does not make much sense to provide a deadline that is in the distant future, nor does it make sense to have a deadline that is before Census Day."

April 5" advisory—printed on the outside envelope of *all* the experimental treatments—is a question that the experiment will not be able to answer.

More fundamentally, experiments and census tests are rarely sized through arguments based on the power needed in support of the statistical tests used to compare alternatives. As noted in Section B–1.c, the critique in our letter report of the lack of a power analysis for the 2010 CPEX experiments—particularly the Deadline Messaging/Compressed Schedule experiment—was answered by the Census Bureau with an appeal to two internal memoranda and an arbitrary doubling of the sample size, with no insight as to how either the original or doubled sample sizes had been derived (U.S. Census Bureau, 2009a). All of this argues for greater attention to standard techniques of statistical experimental design in the planning of census experiments and intercensal tests.

It follows that making improvements in the Bureau's experimental design and testing areas depends on bolstering the technical capability and research leadership of its staff; see Chapter 4 for further discussion of such organizational features.

## Lack of Strategy in Selecting and Specifying Tests, Experiments, and Evaluations

In addition to not providing direct information on cost and quality, the experiments and evaluations in the 2010 CPEX show little sign of anticipating or furthering future methodology that could yield a 2020 census that is reflective of contemporary attitudes, technologies, and available information sources. The choices in the 2010 CPEX seem more suggestive of a "bottom-up" approach—looking at highly specific parts of the census process and making small adjustments—than a more visionary "top-down" approach that takes major improvement in the cost-effectiveness of the census (and such wholesale change in operations as is necessary to achieve that improvement) as a guiding principle.[3] Such a top-down approach would be predicated on alternative visions for the conduct of a census—general directions that might be capable of significant effects on census cost or quality. Then, even though an experiment in the 2010 census would not be capable of fully assessing such a vision, the topics for experimentation would relate to those visions: chosen strategically, they could provide bits of preliminary information to guide later work over the course of the subsequent decade (or decades).

---

[3]Labels like "top-down" or "bottom-up" approaches to planning are, necessarily, oversimplifications. Both approaches—as pure strategies—have value, and the ideal is undoubtedly some combination; we use the labels here to be evocative and to suggest a recent focus on small iterations.

Of the 2000 and 2010 research programs, the only experiments that seem to have taken this kind of strategic approach are the Census 2000 Supplementary Survey (C2SS) and the Administrative Records 2000 (AREX 2000) Experiment, although both of those certainly had limits. The C2SS envisioned the major change of shifting long-form content to the ongoing American Community Survey (ACS). It significantly scaled up the collection of the prototype ACS, although not to a large enough degree to provide extensive information on the estimation challenges that are now awaiting users of 3- and 5-year moving average estimates; the (weak) goal of the C2SS as an experiment was simply to demonstrate that the Bureau can field the decennial census and a large survey simultaneously. The AREX 2000 experiment was a very useful first step in suggesting the use of administrative records in the census process but, arguably, was still focused too heavily on the potential of administrative records as a *replacement* for the census (i.e., do counts and distributions match) rather than administrative records as a *supplement* or an input source to a variety of operations.[4]

Two other points related to the strategy in the selection and execution of census research are worthy of mention. First, research activities are sometimes specified (or misspecified) so that the "next step"—the next key insight or possible outcome—is not taken. For example, the telephone-based Coverage Follow-Up (CFU) operation planned for the 2010 census is a key part of the Census Bureau's coverage improvement activities. The full scope of 2010 CPEX evaluations and assessments relative to CFU is not known to the panel, but based on past Census Bureau history it is virtually certain that the evaluations will provide detail on the number of cases processed in CFU, on the breakdown of cases by incident type (e.g., a household count that conflicts with the number of people reported in the household), and on the number of CFU cases that yielded different responses. However, it is also virtually certain that the telephone-based CFU operation will not include a significant field interview component with a sample of eligible cases; hence, it will not be known how many CFU-eligible cases might have been reached by means other than telephone, nor will it be known how data from the less expensive telephone interviews compare with the "ground truth" established in a face-to-face interview. Likewise, the Census Bureau's two principal geographic resources—the Master Address File (MAF) and the TIGER geographic database—are both examples of cases in which a vibrant research

---

[4]As noted in the synthesis report summarizing the AREX 2000 Experiment (Judson and Bye, 2003:ix), the first objective of the experiment "was to develop and compare two methods for conducting an administrative records census," defined as a "process that relies primarily, but not necessarily exclusively, on administrative records to produce the population count and content of the decennial census short form with a strong focus on apportionment and redistricting requirements." The second objective was to "explore the potential use of administrative records data for some nonresponding or unclassified households."

program should yield regular estimates of geographic accuracy (through random spot-checks and small-scale collection of geographic coordinates). However, current metrics of MAF quality and completeness are generally limited to counts of addresses in the file and rough comparisons with other measures (e.g., independent estimates of the number of housing units).

Second, the Census Bureau has shown an unfortunate tendency to terminate some promising research and development leads too early. To achieve fundamental change, an organization cannot give up on important visions based on initial problems; the Census Bureau's approach is too often to stop at version 1.1 of a promising approach rather than going on to develop 1.2. Arguably, the most prominent example of this tendency in recent experience is the Census Bureau's abandonment of Internet response to the census. Aside from network security and the propagation of "phishing" sites masquerading as the census, the Census Bureau's primary stated reason for its 2006 decision against Internet response in 2010 was less-than-hoped response via the Internet in the 2003 and 2005 omnibus tests. Rather than continue work on constructive ways to bolster awareness of the Internet response options (and acknowledge the shortcomings in design of the 2003 and 2005 tests), the Census Bureau opted to abandon the Internet response option and—worse—to eliminate it from its 2010 CPEX research plans. Another significant casualty of the 2005 test in this regard was alternative structures for the basic residence question (and supporting instructions), including a question-based worksheet approach suggested by the National Research Council (2006) Panel on Residence Rules in the Decennial Census. Based on perceived problems in cognitive testing interviews and less-than-expected performance as one small part of a too-large test, promising ideas in these alternative panels were set aside, when use of a larger number of small, focused experiments could have allowed the approaches to mature.

### Inadequate Attention to the Use of Technology

Past decennial censuses have had to incorporate new technology in important ways. In the 1950–1970 censuses, the gradual shift to a mail-based census and self-response by individuals was also accompanied by development of questionnaires that were machine-readable, reducing the need to key information directly from paper forms. Optical mark recognition (computer parsing of check box and similar information) was complemented in the 2000 census by the use of optical character recognition of handwritten responses; indeed, major pieces of postcensus coverage evaluation work made use of the first-time automated capture of handwritten name information.

Envisioning the development of handheld computers for use in major census operations, the 2010 census promised to make major advances in the use of technology in the U.S. census. In particular, cost savings were projected based on the use of handheld computers in nonresponse follow-up interviewing. The 2010 census is still likely to show technical improvements over its predecessors but—even before the count begins—suffers from the costly and highly publicized breakdown of the handheld computer development. Failures in the Field Data Collection Automation (FDCA) contract between the Census Bureau and the Harris Corporation led to a late "re-plan" of the 2010 census, scaling back of the handhelds to include only the address canvassing operation, an expensive switch back to paper-based NRFU, and a late scramble to complete the operational control systems that govern information flows through the entire census process. The causes of the failure of the full-blown handheld development contract are numerous and beyond the scope of this panel to determine, but we do suggest that a failure to make best use of research and testing played a significant role.

Our predecessor Panel on Research on Future Census Methods reviewed the early plans for the 2010 census and devoted considerable attention to technical infrastructure issues and the incorporation of new technology in the census process (National Research Council, 2004b). The panel recognized that the 2010 census plan included many major system overhauls—not only the development of the handheld computers, but also the establishment of a parallel data system with the American Community Survey and retooling of the Census Bureau's geographic resources. Hence, that panel suggested that:

- Serious institutional commitment was needed to map the logical architecture of the 2000 census, revise that architecture "map" for 2010 census assumptions, and use the resulting model to compare costs of alternative system designs and as a blueprint for final technical systems. In particular, the panel argued that effective systems development would founder without strong "champions" in high management and the establishment of a system architect office to oversee technical development.

- A common pitfall in system redesign is locking into specific physical architectures too quickly. With specific regard to the handheld computers, the panel argued that "the most pressing need regarding [handheld] development is the definition of specifications and requirements—clear statements of exactly what the devices are intended to do" (National Research Council, 2004b:147). Furthermore, the "most important product of [early] testing is . . . a clearly articulated plan of the workflows and information flows that must be satisfied by [the handhelds], as they fit into the broader technical infrastructure of the census" (National Research Council, 2004b:189).

- To facilitate this early focus on requirements, the panel encouraged the Census Bureau to focus more on function than form. "In terms of the capability of the devices likely to be available for 2010, it is almost certain that some testing using high-end devices (e.g., tablet PCs) would provide a more realistic test"—and better sense of requirements—than restricting focus too early on specific palm-size forms (National Research Council, 2004b:147).

On all of these points, the Bureau's development process failed. No system architect position—either for the census as a whole or the handheld computer development in particular—was created, and the logical architecture modeling was little used. In particular, such modeling played no role in the testing of handheld devices in pilot work in 2002 and in the field tests of 2004 and 2006, all of which used devices cobbled together from commercial off-the-shelf components using various palm-size pocket PC-class devices as a base. As described in Section A–7, the 2002 and 2004 activities focused less on requirements of devices than on basic reactions to the devices—for example, would enumerators (with different degrees of experience and familiarity with an area) be comfortable with using the maps on the handheld as a reference? As is now clear from accounts from the need for the census "replan" in 2008, the development process of the handhelds following the award of the FDCA contract (in 2006) was not based on a set of requirements developed from the 2004 and 2006 tests, consequently missing even basic information needs like the need to perform operations in large blocks with hundreds (or thousands) of housing units. Indeed, a final set of requirements for the devices was only developed between November 2007 and January 2008, and the resulting cost estimate from the contractor as to how expensive it would be to meet those requirements precipitated the "replan."

The failures of the handheld development—coupled with the Census Bureau's decision to forbid Internet response to the 2010 census, despite having offered online response (albeit unadvertised) in 2000—have contributed to the strong perception that the Bureau is not adept at incorporating the use of technology. This is an unfortunate situation, because we think it unlikely that cost can be greatly reduced in the 2020 census *without* more effective use of technology.

## Overreliance on the "Special" Nature of the U.S. Census

A factor that looms large in the Census Bureau's decisions to include particular topics in its operational trials and major experiments is whether the topic needs the "census context." That is, the question is whether the topic can best, or only, be tested with full census trappings such as advisories of mandatory response, publicity campaigns, and large sample sizes. To a considerable degree, emphasis on the census context is appropriate because there are features of the U.S. census that make it more than simply a massive household survey. These features include the sheer size and pace of the census enterprise, the reliance of critical procedures on the mobilization of a large corps of temporary enumerators with relatively little training, and the firm constitutional mandate of the decennial count. Still, we think that the Census Bureau frequently exhibits an overreliance on the special nature of the census as it frames its research—a problem that has become part of its culture and attitude toward research.

Put simply, the Bureau's research activities seem premised on the argument that the U.S. census experience is so special—large, complex and unique—that findings can be trusted only if they have been tested in the census context. As we have already noted, our review of the testing and experimentation programs of preceding decennial censuses makes it clear that the Census Bureau used to make much greater use of smaller, focused testing activities, and also used to make greater use of other survey vehicles to test changes that might ultimately be adopted for the census. By comparison, the more recent rounds of census research seem to assume that lessons from small experiments, from general survey research, and from foreign censuses and surveys are somehow inapplicable.

Thus, for example, the Census Bureau determined that it needed to test the effect of sending a second, replacement questionnaire prior to the 2000 census, even though the positive effects of such mailings in general survey research had long been documented. Although the effect of sending a second questionnaire has long been known to boost response rates in general surveys, the Census Bureau determined that it needed to test this extensively prior to the 2000 census. Ultimately, the 2000 census did not include replacement questionnaires because the Census Bureau did not determine the practical requirements of replacement questionnaires until late in the process. There was not sufficient time to work with vendors to accommodate the requirements to print and generate the physical forms in a very short time frame. Although the effect on response rates remained well known, testing the general concept of a replacement questionnaire continued in the 2010 testing round.

A second example of this culture is that, in part, appeal to the special nature of the U.S. census underlies the Bureau's decision not to allow online response in 2010. In this case, the basic argument is that the unique security demands of the U.S. census are such that Internet response in the United States creates too great a vulnerability. While we do not minimize computer security, the implicit argument that those foreign censuses that have implemented Internet response are somehow less or inadequately focused on security discounts efforts by those national statistical offices to ward off hackers and Internet threats.

It is worth noting that threads of the special nature of the U.S. census experience have been part of census culture for a very long time. In fact, we comment in Chapter 1 on what is arguably the first census experiment, the use of advance census forms in the 1910 census. Census Director E. Dana Durand (1910:83–84) described the experiment as "by far the most important method adopted at this census" to increase public awareness of and participation in the census. However, he went on to comment:

> The use of this advance schedule is a partial adoption of the practice of the leading foreign countries in which the larger part of the census work is done by the people themselves, so that the enumerators have little to do in most cases except to distribute and collect the schedules. It is not expected that the same results will be secured by the use of the advance schedule in this country. The novelty of the method, the mixed character of our population, and the complexity of the questions asked—much greater than in foreign censuses—are circumstances which render it likely that a much smaller proportion of the schedules will be properly filled out by families in this country than in countries like England and Germany.

Even from the beginning—the first census after establishment of the permanent Census Bureau—the notion that the complexity of the U.S. census (and population) requires wholly separate tools and methodologies was advanced. Overcoming this insularity—and more effectively building from external researchers and international peers—is a key part of improving Census Bureau research.

## 3–B  KEY STEPS IN RETHINKING THE CENSUS BY RETHINKING RESEARCH

Having critiqued the current state of Census Bureau research, we now turn to suggestions for improvement over the coming decade. We begin by discussing some broad overview strategies before suggesting selected specific ideas with respect to key strategic issues in Section 3–B.4.

### 3–B.1  Identify Visions for Next Census and Focus on a Limited Set of Goals

At the panel's November 2008 meeting, Census Bureau staff discussed a preliminary set of goals and objectives for the 2020 census; they are listed in Box 3-1. It is worth noting that the three labeled "goals" for 2020 in Box 3-1 are essentially identical to those put forward for the 2010 census, save that the Bureau's 2010 goals included a fourth point to "increase the rel-

---

**Box 3-1**   Census Bureau's Tentative Goals and Objectives for the 2020
Census

**Goals**

- Improved accuracy
- Reduced operational risk
- Cost containment

**Objectives**

- Cost savings over repeating the 2010 Census in 2020 (on a per housing unit basis)
- Test and evaluation results incorporated into the design
- Information products and services disseminated electronically that meet customer and stakeholder needs
- Integrated architecture
- Strategic partnerships with stakeholders
- Final core decennial design operations by 2016
- Conduct a full dress rehearsal
- Improved design of data collection instruments
- Improved methodology for enumerating exceptions
- Participative environment for respondents
- Quality assurance integrated into all census operations
- Reduced paper operations
- High response rates
- Uniform coverage
- Reduced duplicates
- Geographic enhancements
- Fully evaluated and researched census
- Coverage measured
- Privacy protected and confidentiality ensured
- An effective and efficient infrastructure

SOURCE: Weinberg (2008).

---

evance and timeliness of census long-form data" through the ACS (Angueira, 2003:2).[5]

We accept this list for what it is—a preliminary first cut—but the first point we make on restructuring census research is related to this listing. The list of objectives contains many good points (although they do sometimes confuse true objectives with the specific tools or procedures intended to achieve those objectives). But it is the sheer length of the list of objectives that we find troubling. In our assessment, organizational success in attaining goals is harmed when the number of objectives being pursued simultaneously is too large. Even a large organization like the Census Bureau, and the

---

[5]The 2010 statement goals also listed the goals in a different order, although no order of importance was explicitly attached to the rankings. The 2010 list (Angueira, 2003:2) lists relevance and timeliness of long-form data first, "reduce operational risk" second, "improve the coverage accuracy of the census" third, and "contain costs" fourth.

management thereof, can focus on only so many large tasks at once. Goals are both useful and necessary; our concern is simply that having too long a list of primary objectives will lead to only incremental progress in meeting any one of them.

We think that a better, research-based path to the 2020 census begins by identifying a small set of alternate "visions" for the 2020 census and then evaluating their possible implications for census cost and quality. "Vision" is necessarily a difficult term to define precisely. By the term we mean a rough articulation of plans for or revisions to each of the major steps of census-taking (e.g., initial contact with respondents, response mode, secondary or follow-up contact with respondents, and information and management infrastructures). The reason for thinking of visions as models for the whole census process is to try to generate ideas that are not so vague or hypothetical as to be unhelpful, yet not so completely worked out as to lock in specific approaches or technologies too early. The "three-legged stool" concept that drove 2010 census planning falls short of a vision in the sense we describe here; although it included revisions of the support infrastructure of the census (geographic resources), it lacked specificity in how the short-form-only census would actually play out. Likewise, the phrase "administrative records census"—in itself—is not really a vision; at least an additional level of detail on how (and how well) administrative data might apply to census operations would be necessary to flesh out the idea and make it a tractable model to consider. Specifically, we recommend:

> *Recommendation 3.1:* **The Census Bureau should immediately develop a limited number of strategic visions for the 2020 census that are likely to meet its announced goals for costs and quality. By strategic visions, we mean start-to-finish strategies for conducting all major census operations in order to confront looming threats and implement new technologies. In addition, evaluation of each vision should include thorough review of the costs and benefits of individual census operations relative to the announced cost and quality goals, to determine whether operations that are not demonstrably cost-effective may be eliminated or scaled back or whether new technologies or practices might usefully be introduced.**

This finite set of visions would provide a starting point for discussion and debate in the early years of the 2010–2020 planning cycle. It is then important that research early in the decade be able to shed light on advantages or disadvantages of the competing visions.

> *Recommendation 3.2:* **The Census Bureau should develop its 2020 census research and development program by identifying a handful of research questions whose resolution will determine**

which of the visions of the next census are feasible and cost-effective and which are not. Priorities for evaluations of the 2010 census should be arrived at consistent with this research program, as should priorities for experiments and tests in the 2011–2018 period.

### 3–B.2  Build Capacity to Evaluate Costs of Alternative Visions

A second major step in a research-based strategy for 2020 relates to the evaluation and comparison of competing visions for the 2020 census. As discussed in Chapter 2, it is remarkable how little is known about the costs of the 2010 census even at this late stage of development. Clearly, the Census Bureau has in place cost models that it uses to develop budget estimates and allocate resources. What is fundamentally unclear is how good those cost models are—how sensitive they are to varying assumptions, how transparent they are in breaking down costs by component operations, and how flexible they are to estimating the costs of major changes to census operations. For example, the Census Bureau acknowledged in October 2009 that—for the address canvassing operation—the Bureau's cost models "did not forecast accurately total costs, and we experienced a cost overrun in components of that operation" (Groves, 2009:7). Specifically, the U.S. Government Accountability Office (2009a:10–11) determined that about $75 million of the $88 million overrun (in an operation originally budgeted for $356 million) was attributable to flawed estimates of workload, both the initial workload for the operation and quality control checks. The remainder of the cost overrun was attributable to the costs of the training and fingerprinting of more temporary staff than needed.

Because we think that census cost and quality are the two central factors that must be addressed in thinking about the 2020 census, it naturally follows that it is of the highest priority that the Census Bureau be able to reliably estimate how much changes in census approaches will affect costs. Accordingly, we recommend:

> *Recommendation 3.3:* **In order to provide early indications of the costs of competing visions for the 2020 census and to support effective planning throughout the decade, the Census Bureau should develop and validate a detailed cost model that not only represents the 2010 census, but also accommodates novel approaches to census-taking, including the use of data capture via the Internet and automated telephone systems, the use of handheld devices in nonresponse follow-up, the use of administrative record information for some types of nonresponse follow-up cases, and innovative mechanisms for reducing the costs of updating the Master Address File between 2010 and 2020. This**

cost model should be able to assess the implications of introducing specific changes to an existing design, singly and in combination, and to distinguish between the direct and indirect cost effects of specific changes. This cost model should be thoroughly documented and transparent so that the Census Bureau can obtain the benefit of expert advice on cost-effective improvements to census operations.

### 3–B.3    Build from 2010 Experience and Data (If Not the 2010 CPEX)

As is clear from the preceding critique in this chapter and the comments in our letter report, it is our assessment that the experiments chosen for inclusion in the 2010 CPEX largely squander a valuable testing opportunity. Save for what might be learned from the small residence piece of the Alternative Questionnaire Experiment (in conjunction with the experimental Group Quarters form; see Appendix B) and the differing number of NRFU contact attempts, the CPEX experiments are not likely to inform changes that will substantially affect either census cost or quality.

We are more optimistic about evaluation work generally (if not the specific evaluation studies currently envisioned in the CPEX framework), contingent on the retention of adequate operational and procedural data as the 2010 census unfolds. Like previous National Research Council panels, we think that a master trace sample that saves and links data for a sample of addresses, respondents, and cases through all steps of census processing would be an invaluable tool for providing empirical insight for intercensal testing. The broader notion advanced by the Panel on Research on Future Census Methods of a master trace *system*—a technical information infrastructure designed in such a way as to automatically and naturally retain virtually all operational data for later reanalysis—is a particularly attractive one.

Due to the 2008 replan of census operations and the resulting crunch to finalize operational control systems (authority for those systems having reverted to the Census Bureau rather than the outside contractor), we recognize that the time and resources to save a designed trace sample simply do not exist. However, as we argue in our letter report, it is absolutely essential that the operational control systems and other census information systems be designed to facilitate data retention—that is, that they include "spigots" or archival outlets to save operational data and facilitate an audit trail. A fully designed sample of cases up front would be ideal, but an archival snapshot of information from all the census information systems in order to build a sample (or trace system) after the census is the next best alternative. Accordingly, we formalize and extend arguments from our letter report as follows:

*Recommendation 3.4:* The Census Bureau should retain sufficient input and output data, properly linked and documented,
from each 2010 census operation to permit adequate evaluation
of the contribution of the operation to census costs and data
quality to feed into 2020 census planning. For this purpose, the
Census Bureau should either establish an internal group or hire
a contractor with database management expertise. This group
would have the responsibility of retaining and documenting sufficient data from the 2010 census to be able to comprehensively
represent the functioning of all census operations. Such a group
would also have the responsibility of assisting Bureau research
staff, using current database management tools, to produce research files to support the assessment of analytic questions concerning aspects of the 2010 census.

We expand on the kinds of data that need to be retained, and the analysis
and linkages that should be explored using them, in discussing options for
the Master Address File in the next section.

### 3–B.4    Examples of Research Directions: Strategic Issues for the 2020 Census

We turn in Chapter 4 to organizational aspects of a successful research
program. First, though, we offer some general comments on four issues that
we think to be particularly strategic concerns for the 2020 census. By this
listing, we do not intend to imply that they are the only important issues
that should go into developing alternative visions for 2020, nor are these
comments meant to be comprehensive treatments of the topics. We merely
suggest that they are sufficiently major issues that some aspects of each of
them should and will pervade such visions; what follows are (admittedly
incomplete) thoughts on possible directions. (One other strategic issue that
rises to this same level—making effective use of testing opportunities in the
American Community Survey—is discussed in Section 4–D.2.)

### Better and Less Expensive Sampling Frame: Directions for the Master Address File

The development and refinement of the sampling frame for the census—
currently the MAF—is clearly a strategic issue for census planning because it
is a key determinant of census coverage. Inclusion or exclusion of addresses
from the MAF has a strong bearing on whether housing units or people are
omitted, duplicated, or misplaced in census returns. It is also a strategic issue
for the decennial census because it is a likely source of hidden costs. More
effective and less duplicative listing could save eventual field costs during

follow-up operations and could reduce the need for broad-brush operations like the complete precensus address canvass used in the 2000 and 2010 censuses. Finally, the effective and accurate upkeep of a MAF is a strategic issue for the Census Bureau—over and above the decennial census—because it is also used as the sampling base for the Bureau's ACS and the major demographic surveys the Bureau conducts on behalf of other federal agencies. Hence, a more accurate MAF at any point in time (not just a census year) benefits most, if not all, of the Bureau's survey programs.

An absolute prerequisite for further research on the MAF and its future improvement is comprehensive evaluation of a type that was impossible in 2000 due to the structure of the file itself. As it existed in 2000, individual list-building operations could overwrite source codes on the address file, so that the most recent operation to "touch" a particular address could be recovered but not the complete history of the presence (or absence) of an address across all operations. Accordingly, a great degree of detective work had to be done to approximate the unique contributions of individual operations to the file and the degree to which operations duplicated each other— detective work complicated by the overlapping schedules of such operations as the Local Update of Census Addresses and the complete block canvass. A major objective of the MAF/TIGER Enhancements Program (MTEP) of the previous decade was to rework and rebuild the format of the database itself; ideally, this has been done in such a way that address source histories are directly recoverable.

Assuming that the source recording in the MAF has been upgraded, then—as part of a master trace sample/system–building effort—an address list research database should be constructed. At a minimum, this research database should link:

- Sufficient "snapshots" of the MAF, with source codes that do not overwrite each other, as to be able to parse out unique contributions of such operations as the twice-yearly U.S. Postal Service Delivery Sequence File (DSF) updates, the Local Update of Census Addresses program (and appeals), the address canvassing operation, and the update-leave operation in 2010, among other sources;

- Information from the 2010 Census Coverage Measurement operations, including whether addresses were flagged as including omitted or duplicated persons (or whole households), nonexistent or nonresidential structures, or erroneously geocoded entries;

- Snapshots of Census Bureau–compiled administrative records databases, such as the Statistical Administrative Records System described below;

- Derived variables about the nature of the housing unit (or structure) at the addresses, such as type of unit (e.g., urban house, rural house, large apartment building, small multiunit apartment building) and demographic area characteristics (e.g., presence in a hard-to-count area due to high prevalence of non-English-speaking households); and

- Returns from the ACS, to add richness (and timeliness) of possible covariates for analysis.

Construction of such a database would be invaluable to sorting out and documenting geographic contributions to census error and the characteristics of addresses that are subject to error. In addition, identifying high degrees of overlap between list-building operations could lead to simplification or consolidation of operations, possibly permitting cost reductions.

In brief, then, some selected issues and possibilities for sampling frame or address list research over the next decade include:

- *Use of address list research database to study feasibility of targeted canvassing operations:* A focus of research should be determining whether it is possible to reliably discriminate between those blocks that are virtually unchanged by the various address building operations (blocks that are stable in that sense) and those that are changed. This could help use the more timely ACS data available in the years prior to the census to steer targeting efforts to the highest priority areas.

- *Quality metrics, change detection, and improved maintenance:* Having made serious investments in upgrading the technical platform of the Bureau's geographic systems during the 2000–2010 decade, the challenge now becomes one of keeping those resources up to date in the most accurate way possible. An original focus of the MTEP was on quality metrics (methods for assessing the quality and geographic accuracy of both the MAF and the line features in the TIGER database) and update mechanisms. One of the CPEX evaluations—comparing the results of detecting whole structures using aerial photography with MAF entries and other sources—may be a useful part of a broader research program in geographic updating. Generally, the Census Bureau would benefit from a program of field spot-checks, comparison with third-party sources (including addresses drawn from administrative records data files), and the like in order to have continuous diagnostic measures of the quality of the MAF and TIGER and to detect priorities for update and maintenance.

- *Continuous address/geographic improvement process:* Several predecessor National Research Council panels on census issues have urged the Census Bureau to make the Local Update of Census Addresses (LUCA) Program a more continuous operation rather than a one-shot (and fairly rushed) chance to review address segments. We concur and urge the Bureau to continue to study the characteristics of addresses added or deleted by LUCA partners, and we also urge the Bureau to consider a broader approach: a more continuous local geographic partnership for both the MAF and the TIGER database. Means of combining opportunities for local review of portions of the MAF with the regular Boundary and Annexation Survey used to update political boundaries, further combined with periodic sharing of locally maintained geographic information system files such as were a major source of information in the main TIGER realignment project of the past decade, would benefit both the Census Bureau and state and local governments. Developing ways in which continuous geographic updating can be made easier for local government participants—for instance, through software interfaces that make it easier for governments to respond using their existing electronic files or through consistent use of identifier codes for MAF and TIGER features—would serve to bolster participation by a wider range of governments. A continuous program of geographic resource improvements should also be accompanied by the development of quality and coverage metrics for both the MAF and TIGER (discussed above), so that the quality and unique contribution of local update sources can be assessed and areas with particular need for updating can be identified.

- *Integration with the American Community Survey field staff:* Another original plank of the MTEP was what was known as the Community Address Updating System (CAUS)—effectively, making use of field staff assigned to ACS collection to make geographic updates if they encountered new addresses or streets on their rounds. CAUS failed to emerge into a major presence in recent years because of funding constraints and, perhaps more fundamentally, the more pressing exigency of simply getting the ACS on a solid footing. Still, the concept is potentially sound and useful. One possibility that could be researched would be periodic, systematic additions of address list or map segment verification tasks to ACS interviewer workloads, rather than simply enabling geographic updates if interviewers happen to come across new addresses or developments while on their rounds. Possibilities for study might include an approach known as half-open intervals: directing interviewers to go in some direction from a household in their interviewing workload and list any addresses they cannot find in MAF entries, stopping at the first unit they encounter that is on the MAF. Use of ACS field staff for such geographic update activities is certainly not a perfect solution for MAF updates over the course of a decade. However, this work could help in quality measurement of MAF and TIGER and provide clues to detect areas where the Bureau's geographic resources might require particular updating.

- *Position the MAF as a national resource:* The MAF benefits other
  federal government agencies through the major demographic surveys
  (done under contract with the Census Bureau) that use the MAF as a
  sampling frame. It is also developed, in some respects, in partnership
  with the U.S. Postal Service because the postal DSFs are a major input source to the MAF. It stands to reason, then, that a useful area of
  research concerning the MAF would be whether it satisfies the needs
  of its major stakeholders and insights that other agencies may have on
  the frame-building process. In the case of the Postal Service, our review of past decades' research programs—replete with intensive use
  of postal checks and use of information collected directly from local
  letter carriers—is a reminder that establishing and maintaining a research partnership with the Postal Service is vital. For example, it
  should be determined whether the Census Bureau's adaptation of the
  regular DSF updates makes use of the full range of address information on those databases and whether other postal data could further
  improve the MAF or TIGER. It should be noted that broader use of
  the MAF would be likely to require action by Congress (akin to the
  1994 act that permitted both LUCA and DSF updating) because of the
  U.S. Supreme Court's interpretation that Census Bureau address lists
  fall under the confidentiality provisions of Title 13 of the U.S. Code
  (*Baldrige v. Shapiro*, 455 U.S. 345, 1982).

- *Interface with the commercial sector:* Just as it is important to compare the address coverage from compiled administrative records files
  with the existing MAF, it would also be worthwhile to study the quality and coverage of commercially available mailing lists (even if such
  commercial lists do not become an input source to the MAF). In particular, it would be useful to learn from private-sector practices in building, maintaining, and filtering mailing lists, as well as how private-
  sector firms have developed other frames, such as e-mail and telephone
  listings.

- *Assess integration of the "household" MAF with group quarters:* One
  of the improvements promised in the MTEP was the merger of MAF,
  TIGER, and group quarters information into a common database
  structure. For the 2010 census, group quarters validation will still
  be done as a separate operation, following up with structures and facilities labeled as "other living quarters" in the complete address canvassing operation. Study of group quarters validation will be useful for
  judging the effectiveness of the merger of these lists and the ability of
  flags in the address records to distinguish between group quarters and
  regular household populations. This kind of study and research is particularly important given the sometimes blurred line between conven-

tional household and group quarters structures, such as group homes and health care facilities that may combine "outpatient," independent living, assisted living, or full-time nursing care functions within the same structures.

## Better and Less Expensive Data Collection: Toward a "Paperless" Census

An original goal of the incorporation of new technology into the census process for 2010 was to reduce the use, storage, and movement of paper in the census. The use of paper translates directly to both cost and time; while the use of document scanning greatly reduces the time that needs to be spent handling the paper (keying information directly from forms), the reliance on paper at all stages has serious implications for the size and scope of the local census offices and data capture centers that must be equipped for the census. The use of technology is an area in which the 2010 census is likely to be remembered for some strides but probably more for the costly and embarrassing collapse of the plans for use of handheld computers in NRFU interviewing and reversion to paper-and-pencil methods.

That the 2010 census will fall short of its original goals for reducing paper use is not a failure of vision—the goal was a good and laudable one—but a failure to execute and fully articulate that vision. The idea itself was not enough to guarantee success: the idea had to be matched by a set of research and testing activities designed to propel the larger task of technology development forward and specify the requirements for the technology.

Going forward, it is difficult to imagine a plan for the 2020 census that can substantially reduce costs or increase quality without a major emphasis on developing and integrating new technology. As a bold statement of mission, we encourage the Census Bureau to go further than to think of simply getting the development process of handheld computers for NRFU in shape. Rather, we suggest a broader examination of all steps in the census process with the public, stated goal of making the 2020 census as "paperless" as is practicable.

Further reasons why an effort to move the census in a paperless direction is a critical strategic issue for 2020 include the implications for quality. Indeed, we think that experience in the general survey research community suggests that the gains in accuracy from electronic methods for data collection may be more important than cost reductions. The Census Bureau's own work in the 2003 and 2005 tests suggested that Internet responses were typically of higher quality than responses by other modes; edit routines and skip patterns in electronic questionnaires can promote accuracy in subtle but important ways. Secondary gains in accuracy are not hard to imagine. Nonpaper formats would not have the same hard space limits as paper forms, thus reducing the need for follow-up with large households for whom reported information simply does not fit on the paper form. Questionnaires could also be made directly accessible in a wide variety of foreign languages, without the strong filter of recent censuses in which a call to an assistance center was necessary to request a foreign language form. The age distribution and attitudes of the population also make a higher tech, relatively paperless census a key strategic issue; new generations are arguably more conversant with electronic media than paper media, and a "green" census (saving paper) might serve as a strong incentive to boost participation. However, one of the strongest arguments for a heightened focus on use of technology leading to the 2020 census is simple perception, exactly the reason why the 2010 census looks odd relative to other national censuses and surveys that are now turning toward Internet data collection. That is, it would simply look foolish and out of step to try to force 2020 census technology into the 2010 mold rather than aggressively studying and rebuilding systems.

The guidance by the Panel on Research on Future Census Methods in its final report (National Research Council, 2004b) on developing and implementing a technical infrastructure remains valid. It also follows that movement toward census processes that are highly automated and as paperless as possible heightens the importance of ensuring that those processes have an audit trail—that they include outlets for retention, archival, and analysis of operational data such as we recommend for the 2010 census in Section 3–B.3. Having already described many of the points raised in that report, we do not expound on them further. In brief, some other selected issues and possibilities for technology research over the next decade include:

- *Boosting response using paperless methods:* One of the valid arguments raised by the Census Bureau for not permitting online response in the 2010 census is that their experience suggests that permitting electronic response options along with a paper questionnaire does not seem to elevate overall response. That is, it does not seem to produce original respondents: it may sway some people who might otherwise respond by mail to return the form electronically, but it does not convert probable nonrespondents to respondents. This observation is not unique to the Census Bureau's experience; other national statistical offices and survey research organizations have encountered the same phenomenon. Developments in questionnaire design and approach strategies should be pursued by the Bureau in cooperation with these other groups.

- *Security and confidentiality concerns:* In overcoming the concerns about computer and network security that led it to disallow online response in 2010, the Census Bureau would benefit from in-depth study of the security mechanisms used in other censuses and surveys. It would also benefit from examples in the electronic implementation of other government forms, such as tax returns.

- *Mode effects:* A perennial concern in survey methodology with the adoption of new survey response types is the difference in consistency of response across the different types. Response mode differences are not inherently good or bad but need to be understood and documented.

## Better and Less Expensive Secondary Contact: Nonresponse Follow-Up

Reexamining assumptions and strategies for NRFU operations is a key strategic operation because of the significant costs of mobilizing the massive temporary enumerator corps and making contacts at households that, for whatever reason, do not respond to lower cost initial contacts.

Research on *the possible role of administrative records in NRFU processes* is particularly critical to achieving better and less expensive secondary contact with respondents. With some additional evaluative and follow-up components, the telephone-based 2010 CFU operation could provide some useful insight to start such research. One of the possible planned sources of household or address records being submitted to the CFU operation is a search of census returns against the Census Bureau's database of administrative records compiled from other federal agencies, a database currently known as StARS (Statistical Administrative Records System) or e-StARS. As we have also described, the retention of operational and procedural data during the 2010 census also has the potential to yield very valuable information; these data snapshots should be able to support a post hoc examination—as a research question—of the impact on census costs and quality if administrative records had been used as supplemental enumerations at various stages of NRFU work. All stages—from near-original enumeration of nonresponding households to use of records as a last resort measure rather than proxy information—should be considered and investigated.

To be clear, the use of administrative records should not be seen as a panacea for all census problems, and we do not cast it as such. Sheer numeric counts aside, the quality and timeliness of administrative records data for even short-form data items, such as race and Hispanic origin and relationship within households, remain open and important questions. Wider use of administrative records in the census also faces formidable legal hurdles, not the least of which are inherent conflicts between the confidentiality and data access provisions in census law (Title 13 of the U.S. Code) and the Internal Revenue Code (Title 26), given the prominence of tax return data in the administrative files. Still, just as it is difficult to imagine a 2020 planning effort that seriously addresses cost and quality issues without aggressive planning for use and testing of new technology, it is also difficult to imagine such an effort without a meaningful examination of the role of administrative records.

Other key strategic issues for NRFU-related research include:

- *Investigation of state and local administrative data sources:* As mentioned above, the Census Bureau's current StARS administrative records database is built annually from major administrative data sources maintained by federal agencies, including the Internal Revenue Service. Particularly as the idea of using administrative records in a variety of census operations (such as geographic resource updates) is considered, the Census Bureau should explore the quality and availability of data files maintained by state and local governments, including local property files, records for "E-911" conversion from rural non-city-style addresses to easier-to-locate addresses, and state and county assessors' offices.

- *Optimal pacing and timing of contacts:* The NRFU Contact Strategy Experiment of the 2010 CPEX varies the number of contacts allowed for nonresponding households. As we have already noted, there is something slightly off in the specification of the experiment—capping the number of visits and making this known to the interviewers presents opportunities for gaming the system. But the optimal number of attempted NRFU contacts—based on yields of completed interviews and quality of information—is an important parameter to resolve. So, too, is work on best ways to structure local census office and enumerator workload in order to maximize the chances of successful interview completion.

- *Efficacy of telephone follow-up:* Along the same lines, a research question that has been touched on in past census research but that is worth revisiting in the 2020 climate is use of telephone (or electronic) follow-up rather than personal visit. The effectiveness of the telephone-based CFU operation in 2010 may provide initial insight on the feasibility of conducting such operations on an even larger scale. Much may also be learned about the effectiveness of telephone-based follow-up in the census context by studying and evaluating its use in the ACS and other Census Bureau surveys.

- *Reducing NRFU workload by shifting some burden:* Clearly, a critical determinant of the cost of NRFU operations is the number of nonresponding households that are followed up. The U.S. Supreme Court's 1999 decision and current wording of census law reinforce that reducing that workload by following up with only a sample of households is not permissible. But research efforts on another angle to cut into the overall NRFU workload—promoting late response to the census—may be worthwhile. This could involve, for example, extending publicity and outreach campaigns (and possibly some shift in message—emphasizing that "it isn't too late" to respond) and "Be Counted"–type programs; such a message was employed in the 1980 census (see Table 2-5). The effectiveness of such an approach would depend on the level of automation of census operations (e.g., the ability to transmit quickly revised enumerator assignments) and the time demands for data capture from paper forms. Still, the costs and benefits are worth exploring—these efforts might not sway some truly hard-to-count respondents, but they could elicit responses from some reluctant or forgetful households.

- *Examining relative quality of "last resort" enumeration:* In those cases in which contact simply cannot be made, the relative quality of different options for filling the blanks—for example, proxy information, imputation, and use of administrative records—should be quantified and evaluated.

- *Quality of interviews as a function of time from Census Day:* It is generally well understood that follow-up interviews (as well as independent interviews such as the postenumeration survey that is the heart of coverage measurement operations) are best done as close as possible to the census reference date. Doing so helps to curb discrepancies due to people moving to different households or switching between "permanent" and seasonal residences. It is also generally well understood that there is decay in interview quality and consistency with length of time from the survey. This is arguably more an issue for time-sensitive information (i.e., exact knowledge of monthly utility bills) or recall of numerous events than for the items on a short-form-only census. Still, a body of quantitative evidence on recall and decay effects on short-form items (including number of persons in and composition of the household)—and key long-form items currently collected on the American Community Survey—as they vary with time from Census Day would be very useful in revisiting such assumptions as optimal timing of determining the start of NRFU operations.

## Rethinking the Basic Census Objective: Getting Residence Right

The basic constitutional mandate of the decennial census is to provide an accurate resident count. Accordingly, in terms of setting basic strategy for the 2020 census, the concept of residence and collecting accurate residence information is vitally important. Census residence rules and concepts have been easy to pigeonhole as a questionnaire design issue—the search for the right phrasing and ordering of words and instructions at the start of the census form in order to prod respondents to follow particular concepts. These questionnaire design matters are important, but the issues are much broader—a thorough examination of residence merits attention to the basic unit of analysis of the census, to the implications of residence concepts for data processing design of operations, and to tailoring enumeration approaches to different levels of attachment to a single "usual" place of residence.

The National Research Council (2006) Panel on Residence Rules in the Decennial Census discussed a wide range of research ideas under the general heading of residence; these generally remain as applicable to the 2010–2020 planning period as they were to 2000–2010. In terms of questionnaire design, these include further research on replacing the current instruction-heavy approach to the basic household count question with a set of smaller, more intuitive questions and more effective presentation of the rationale for the census and specific questions. Other specific research suggestions (many of which draw from the same panel's recommendations) include:

- *Quality of facility records for group quarters enumeration:* Residence concepts and group quarters enumeration are impossible to disentangle because the nonhousehold, group quarters population includes major cases in which determination of a single "usual" residence is difficult: college students away from home, persons under correctional supervision, persons in nursing or other health care facilities, and so on. Hence, attention to getting residence concepts right demands attention to methods for enumerating group quarters. In the 2000 census, about half of the returns for the group quarters population were filled through reference to administrative or facility records rather than direct interview. Yet much remains unknown about the accuracy and capability of facility records and data systems to fill even the short-form data items, let alone their ability to provide the kind of alternative residence information that would be necessary to inform analysis of census duplicates and omissions. The National Research Council (2006:240) suggested that the Census Bureau study the data systems and records of group quarters facilities on a continuous basis, akin to its efforts to continuously update the MAF. This is clearly a major endeavor, but one that is particularly important because of the inclusion of group quarters in the ACS. Even the records systems of state-level correctional systems will vary, let alone the plethora of systems maintained by individual colleges or health care facilities. But research toward a continuous inventory would benefit other surveys of group quarters populations between censuses and may suggest methods for more efficient and accurate data collection from the relatively small but very policy-relevant group quarters population.

- *Residence standard mismatch between the census and the ACS:* While the decennial census uses a de jure "usual residence" concept, the long-form-replacement ACS uses a 2-month rule—effectively, a de facto or "current residence" standard. The exact ramifications of this difference in residence standards are unknown, and, indeed, they may be relatively slight, particularly given the pooling of ACS data to produce multiyear average estimates. But more precise empirical understanding of the possible differences introduced by differing residence standards in the Census Bureau's flagship products would bolster the ACS's credibility. In the next chapter (and in Recommendation 4.5 in particular) we discuss the need for integration of research between the decennial census and the ACS; a matching study of the census to ACS returns near April 2010 would be an ideal step in that regard to study possible differences in residence specifications and household rostering.

- *Revamping "service-based enumeration":* In the 2000 census, the Census Bureau's efforts to count transient populations, including persons experiencing homelessness, were combined into an operation known as service-based enumeration. This operation—to be repeated in 2010—relies principally on contacts at locally provided lists of shelters and facilities providing food or temporary shelter services. Just as group quarters enumeration would benefit from sustained research effort and attention over the decade, so, too, would outreach efforts to best cover the service-based population. Such effort should include collaboration with subject-matter experts, local government authorities, private service providers, and other agencies (such as the U.S. Department of Housing and Urban Development) that have periodically attempted measures of the number of homeless persons at points in time. It should also include focused, relatively small surveys to compare the efficacy of sample-based measures of the homeless population compared with census-type canvasses.

- *Revisiting the "resident" count:* Because of the legal climate surrounding the 2010 census, the Census Bureau may face pressure to conduct research on components that are currently included or not included in the census "resident" count. It should prepare accordingly. In particular, its research program should give some thought to studying the effects on response and cooperation by including questions on citizenship or immigration status. The arguments that such questions could seriously dampen response and hurt the image of the decennial census as an objective operation are straightforward to make and, we think, are basically compelling, but empirical evidence is important to building the case. With regard to counting Americans overseas, the experience of the 2004 Overseas Enumeration Test is very useful, but here, too, additional quantitative evidence would be useful. In particular, it would be useful to examine and critique information resources that may be available from the U.S. Department of State (e.g., contacts with overseas embassies or consulates) to estimate the level of coverage in such files; it would also be useful to evaluate and assess the quality and timeliness of the data files that the Census Bureau already uses from the U.S. Department of Defense and other federal agencies with employees stationed overseas (for inclusion in apportionment totals).

## Planning Now for the Census Beyond 2020

The history of past census research that we have outlined in Appendix A and described in this chapter—particularly the successful adoption of mailout-mailback methods—suggests that truly massive change in approach to the census can take decades of planned research to be fully realized. It is vitally important in 2009 and 2010 for the Census Bureau to be thinking of and planning for the 2020 census, but it is also appropriate to be thinking now of even broader changes that may apply even further in the future.

A sample of such broader, fundamental issues that should be considered for long-term research efforts include the following:

- *A census without mail:* Arguably the boldest area for research in this direction is the concept of a census *without* primary reliance on mailout-mailback methods. Given the difficult fiscal circumstances of the U.S. Postal Service and major effects that electronic commerce and e-mail have had on regular physical mail volume, means for making initial contact with the national population other than mailed letters or questionnaires may have to be considered in future censuses.

- *Change to the unit of enumeration:* Since the act authorizing the 1790 census required the count to be counted at their "usual place of abode," the decennial census has used the household as its basic unit of enumeration. In the modern census context, this has involved associating households with specific addresses and, through those addresses, with specific geographic locations. Just as the core assumption of mailout-mailback methodology is one that should be probed and reconsidered in coming years, so too is the unit of enumeration worthy of research and examination. For example, it is important to consider how a census using the individual person as the unit of analysis can be analyzed and tabulated, as well as the extent to which households, families, or larger constructs can be reconstructed from a person-level census.

- *Interaction between the census and the American Community Survey:* We discuss the integration of the census and the ACS further in Chapter 4, but the topic is a critical long-term research enterprise. In its early period of development, it is both appropriate and important to focus on the properties of the ACS as a replacement for the long-form sample of previous censuses—whether ACS tabulations can satisfy both user needs and the myriad legal and regulatory demands for demographic data. Going forward, the capacity of the ACS as a unique survey platform in its own right must be explored, including ways for the census and the ACS to support each other: for example, use of parts of the ACS sample as a test bed for experimental census concept and questions.

# – 4 –

# Revitalizing Census Research and Development

THE CENSUS BUREAU NOT LONG AGO led the world in goal-oriented research and development (R&D) for continuous improvement of its censuses and surveys. The fruits of that R&D included such path-breaking achievements as:

- the use of probability sampling in censuses and surveys (first used in the decennial census in 1940), which dramatically reduced respondent burden and the costs of data collection compared with a complete census, while allowing the collection of detailed information with known error due to sampling;

- computerized processing of census returns, begun on a small scale in the 1950 census and fully implemented in the 1960 census, which made it possible to deliver detailed census results on a faster schedule, improve methods for handling missing data by using "hot decks" instead of "cold decks," and dramatically increase the data products provided to users, including public-use microdata samples, first produced from the 1960 census in 1963;

- mailout-mailback enumeration, partially implemented in the 1960 census (the mailout portion) and fully implemented for much of the country in the 1970 census, which reduced errors in coverage and content (self-reports on census mail questionnaires are more accurate than enumerator reports) and, at least initially, reduced the size of the enumerator workforce;

- the use of dual-system estimation for census coverage measurement, first implemented in the 1980 census, which made possible more accurate estimation of net undercount by a "do it again, independently" approach, compared with the "do it again, better" approach used in the 1950 and 1960 censuses, in which enumerators rechecked the counts of housing units and people in sampled areas; and

- the TIGER geographic coding and mapping system, developed for the 1990 census, which made it possible for the first time to generate maps and geocode addresses by using a computerized database that represented physical features, census geography, and street networks for the entire country.

More recently, the Census Bureau has successfully designed and implemented the American Community Survey (ACS) as a replacement for the census long-form sample. And the Census Bureau has many innovations to its credit in other programs, such as its economic censuses and surveys and its household surveys.

Yet over the past two or three decades, there has been significant erosion in the Census Bureau's once preeminent position as a world leader in statistical research and development. The cumulative effects of actions and inactions—on the part not only of the Census Bureau, but also of the Department of Commerce and Congress—have led to a situation in which research and development for the decennial census and other programs too often is limited to incremental improvements in existing systems, is planned from the bottom up without sustained top-down strategic direction, is executed without the benefit of using best practices for the design of experiments and tests, expends scarce resources on testing factors that are already well established in the literature while neglecting to test factors that are unique to the scope and scale of the census or another program, is fragmented organizationally, is not well integrated with operations, is not considered a key driver of future directions or new operational procedures, and lacks resources commensurate with needs.

The results of an inadequate and unfocused research infrastructure for the decennial census are evident in the failure to carry out the planned development of handheld technology for nonresponse follow-up in the 2000 census, the failure—even after several decades of on again, off again effort—to make significant use of administrative records in the census and household surveys, the failure to use the Internet in the 2010 census or in household surveys (a test of an Internet response option is planned for the ACS), the failure to adequately evaluate and improve the procedures for updating the Master Address File, the limited and unfocused experiments planned for the 2010 census, and the lack of clearly specified "stretch" goals for planning the 2020 census that are designed to break the unsustainable trend of escalating costs and complexity of census operations.

In this chapter, we not only describe the functions and properties of an effective R&D program for a major statistical agency in general terms, but also make specific recommendations to revitalize the R&D function at the Census Bureau. Given our charge, we focus on R&D for the decennial census, although many of our comments may apply to R&D for other bureau programs as well. Section 4–A begins by fleshing out what we mean by R&D in the context of a statistical agency followed by a description in Section 4–B of the properties of a successful R&D program for the Census Bureau. We then turn our attention to the organizational structures around R&D (4–C) before closing in Section 4–D with recommendations for developing an improved census R&D environment.

## 4–A  IN-HOUSE R&D—WHY AND WHAT

We begin by dismissing any thought that a statistical agency, such as the Census Bureau, does not require a significant in-house R&D capability. R&D is central to the ability of a statistical agency to carry out its mission to deliver relevant, accurate, and timely statistics to the public and policy makers in the face of changing data needs that reflect a changing society, declining public cooperation with censuses and surveys, constrained staff and budget resources, and changing technology for data collection, processing, estimation, and dissemination. The Committee on National Statistics in its *Principles and Practices for a Federal Statistical Agency* (National Research Council, 2009b:11–12, 43–45) specifies an "active research program," including substantive analysis and research on methodology and operations, as 1 of 11 essential practices for a statistical agency. Indeed, unless an agency is simply a data collection contractor to other agencies that provide the ongoing scientifically based leadership for censuses and surveys, then it must itself have an ongoing, high-quality, adequately resourced in-house R&D capability. Even for those surveys in which the Census Bureau is the data collection contractor, it behooves the Bureau to continually improve all of its statistical capabilities, such as sampling, editing, quality assurance, data collection, data processing, software development, and analytic approaches, similar to what the major private-sector survey data contractors do in their efforts to be competitive and provide customer value.

There are a number of ways to define R&D, including the classic distinctions of "basic research," "applied research," and "development," which actually work well for our discussion. We define R&D to include the following components:

- "Basic research," by which we mean analytical work that is ongoing and devoted to fundamental problems of improving relevance, accuracy, timeliness, and efficiency of a statistical agency's data programs. Such research might, for example, investigate alternative methods for imputing missing responses, including not only the traditional hot-deck method, but also model-based multiple imputation, in a wide variety of survey contexts. Or such research might investigate ways to improve the timeliness and accuracy of census and survey response through redesign of questionnaires in a variety of modes, including mixed-mode census and survey designs.

- "Applied research," or "applied methods," by which we mean analytical work that is directed to the specific needs of a specific census or survey program. Such work would take research findings and adapt them to a specific context by, for example, providing weighting or imputation specifications for a particular census or survey.

- "Development," by which we mean work, involving some combination of researchers, methodologists, and operations people, to implement research findings on the necessary scale for a census or survey. Some of the recent failed attempts to reengineer the decennial census, such as the collapse of the plan to use handheld technology for nonresponse follow-up, have involved a failure to conduct the needed developmental work with sufficient lead time.

While we strive to make clear when we are talking about one of the three components listed above, we also use "research" as a short hand for the entire array of activities that must be part of a statistical agency's R&D portfolio in order to ensure that its data are as relevant, accurate, and timely as possible within resource constraints.

## 4–B   PROPERTIES OF A SUCCESSFUL R&D PROGRAM

To be successful, a research and development program for a major statistical agency of the size and scope of the Census Bureau should have the following characteristics:

- *Research activities related to strategic goals and objectives:* In the case of the decennial census, the overarching goals of methodological research and development are to materially reduce costs and increase (or at least maintain) quality in terms of the coverage of the population and the completeness and accuracy of responses to content items. Therefore, all R&D projects should be justified on that basis. Furthermore, each cycle of census design work needs to start with the development of a small number of competing visions for the next census, in which the ultimate selection of the vision to use as the foundation for the design of the next census depends on the resolution of a handful of basic research questions. Any research that helps to address these fundamental questions should be given a higher priority than research that is not associated with those questions.

- *Research-supported decision making:* There is evidence that some of the major census innovations, or attempts at innovation, have been implemented without sufficient support from census experiments or tests. Examples include the inadequate testing of the census handhelds in the 2010 planning cycle, the inadequate testing of the optical scanning procedure in the 2000 cycle (which nearly resulted in a major delay for the 2000 census data collection effort—see U.S. General Accounting Office, 2000), and the inadequate operational testing (as preparation for implementation) of the use of a targeted replacement questionnaire leading up to the 2000 census. Research needs to be seen as an initial, key step in all major decisions concerning decennial census design. Accordingly, the outputs, or evaluation metrics, for each research project need to be carefully specified—for example, whether a particular test of a handheld device for census-taking is primarily to assess data quality or operational feasibility or costs or some combination—and provision made to collect the necessary information in a form that can readily be analyzed.

- *Appropriate balance between fundamental research and applied methodology:* The research program at the Census Bureau needs to emphasize basic studies aimed at establishing general principles for the design of censuses and surveys as much as, if not more than, it emphasizes applied studies designed to determine how these principles apply to specific surveys. Thus, research on the census that is too context-dependent and too focused on the immediately upcoming census will probably not yield results that are helpful for the next census, with the consequence that the R&D cycle for that next census will have to start afresh with little cumulative knowledge gained from prior research. Moreover, while the decennial census is relatively singular in such features as its large scale, extent of public scrutiny, and unforgiving timetable, there are important commonalities between the census and other household surveys, in particular the ACS. Consequently, research that addresses fundamental issues—such as why certain types of question formats or certain data collection modes elicit more or less complete and accurate responses—is more likely to yield results that help more than one census or survey in more than one time period than is research that is too specific to a particular survey and time period.

- *Continuity over the decades:* Successive stages of research on a given topic need to build on previous results, otherwise they are reinventing the wheel, or else the resulting disparate research findings from isolated tests and experiments will be difficult to evaluate and connect to existing theory. Each successive research activity needs to incorporate what was learned in previous research activities about the question at hand through the choice of appropriate control treatments, alternative procedures, and environments of study. For example, a postcensus questionnaire test should include as a control the previous census questionnaire or, alternatively, a questionnaire that was tested in that census and proved efficacious (this was not done in the 2010 questionnaire testing conducted in 2003). Moreover, substantial development research and testing followed by operational testing will generally be needed for innovations in decennial census design given the heterogeneity of the U.S. population, its living situations, and questions of scale. Ideally, such research would build on work conducted for the previous census and the ACS.

- *Adequate expertise and professional development:* Research should be seen as having a very high importance in the organization, and this would be evident in the size and funding of the research group, the talent of the staff, and their role in decision making. An effective research program for the census and surveys would have staffing—with many personnel at the doctorate level—with expertise in experimental design, survey design, the technology of survey data collection, cognitive methods in survey research (especially questionnaire design), geographic information systems, database management tools, and statistical methods in such areas as record linkage, analysis of complex survey data, survey variance estimation, and methods for treatment of missing data. Such staff should have adequate support to not only maintain, but also continually develop their human capital—for example, by being funded to attend several technical conferences a year and encouraged to prepare research papers for publication. The research staff should be afforded opportunities for direct and frequent interaction in teams with Census Bureau field and program staff across the Bureau's organizational divisions. In addition, the research staff should have the capability for regular interaction with external experts through not only advisory committees, but also appropriate contracting mechanisms that provide for more extended interaction. The ability to work directly with external experts is critical to enable the in-house research staff to keep abreast of innovations in survey methodology in academia and the major private survey research corporations.

- *Information technology development:* Another important area of expertise for the Census Bureau's research staff should be information technology (IT) knowledge and skills that permit the staff to work effectively with IT contractors. If it is difficult to attract a sufficient core of in-house staff with expertise in systems and software design, then it becomes even more important to reach out to academic and private-sector experts who can function as part of the in-house group and provide valuable guidance on such matters as evaluating proposals from contractors and overseeing the work on major IT contracts. Undoubtedly contributing to the Census Bureau's failure to successfully manage the contract for use of handheld computing devices in the 2010 census was the lack of integration of the contractor staff with in-house technical staff.

- *Consistent use of state-of-the-art experimental design methods:* The research group should identify and follow sound principles and practices for the design of experiments and tests and update them as the state of the art advances; Chapter 3 discusses current deficiencies along these lines in more detail.

  Fundamentally, as we discuss in Sections 3–B and B–1.c in this report and in our interim and letter reports, census experiments and tests are rarely sized through explicit estimation of the power needed to support the statistical tests that will be used to compare the effects of alternative treatments. An undersized test, in terms of the number of completed sample cases, will not permit conclusive analysis of the effects of one or another treatment. Relatedly, testing resources are often wasted by the failure to target relevant population groups. For example, if tests of variation in question wording for eliciting responses from small ethnicity groups, such as Afro-Caribbean, are not targeted to areas of expected concentrations of such respondents, the tests are likely to collect too few cases of interest while at the same time wasting taxpayer resources and respondents' time by collecting a large number of irrelevant cases.

- *Appropriate balance of types of research and testing:* The R&D cycle in recent censuses has focused on large-scale tests, such as a complete census operation in a locality or mailings of thousands of questionnaires to test wording alternatives. The only inputs to large-scale questionnaire tests have generally been cognitive testing with very small numbers of respondents (fewer than 10 people). While both large- and very small-scale tests have their place, it is important for decennial census R&D to include other research and testing methods. Targeting of questionnaire tests, for example, could reduce the number of respondents required and thus make better use of scarce resources. A series

of smaller tests focused on potential new features of census-taking—
for example, a series of Internet data collection tests—would probably
be more cost-effective than two large tests, the first of which does
not typically provide results in time to affect the second test. In addition, cost-effective cumulative R&D for the census would make extensive use of such techniques as simulation of changes in operations—
such as a targeted rather than complete address canvass—using well-
documented databases from the previous census. Relatedly, wherever
possible, the "not-invented-here syndrome" would be rejected in favor of adopting well-established methods from other organizations.
For example, as mentioned above, the Census Bureau conducted an
elaborate line of testing of multiple questionnaire mailings in the early
1990s. This work was solid and demonstrated gains in response rates
that could be achieved through replacement questionnaire mailing;
however, to a large extent, it replicated work on mailing package research and confirmed findings that were already known in the survey
research literature. The consequence was that relatively little had been
done on developmental work—developing operational specifications
to determine whether multiple mailings were feasible on the scale and
timetable required for the census—until it was determined that a second questionnaire mailing could not be successfully used in the 2000
census.

- *Facilitated access to data outputs:* Data on the outcomes of experiments, tests, and other research—such as effects on response rates
  or the distribution of imputations or the costs of operations—should
  be made available to the research group in a form that facilitates exploratory and confirmatory analysis. More concretely, this means that
  research projects should produce outputs that are well documented
  and provided in databases that are easy to access for a wide range
  of different analysis, using different covariates and statistical measures. It also means that operational tests and, indeed, full-scale census
  operations—for example, nonresponse follow-up or data capture—
  should record and store transactions in well-documented formats that
  researchers can readily access for cost-modeling or evaluating the effects of one or another operation on data quality.

- *Research on implementation and human factors:* There is a role in
  census research for small-scale tests or experiments of potential innovations in methodology, just as there is a need for research that
  establishes the feasibility of those innovations at a census scale of operations. The trick lies in balancing these activities and not—as in previous recent censuses—favoring complete tests of all census operations
  in one or more locations to the exclusion of smaller, focused tests that

could have been more efficient and effective. Useful, midlevel research between these extremes could involve working with vendors and Census Bureau field division staff to identify requirements to bring innovations to scale and to conduct tests of specific components to determine operational feasibility. An important aspect of feasibility testing should be the explicit consideration of human factors, such as whether an innovation alters the division of responsibilities among enumerators, local census offices, regional offices, and census headquarters and the flows of information among them in productive or counterproductive ways. Although the planned use of handheld devices for nonresponse follow-up had major implications for the interactions of enumerators, local and regional offices, and census headquarters, such human factors were not explicitly part of the testing program.

## 4–C  STRUCTURING A SUCCESSFUL R&D PROGRAM

The conduct of relevant, high-quality, and timely censuses and surveys within resource constraints is a complex enterprise, which depends on research that is integrated into, yet independent from, daily practice. The success of an effective, well-integrated research program depends critically on the Census Bureau's structure for research and how leadership and organization permit research to interact with, and not be impeded by, the constraints of census operations.

Unfortunately, there is no pat answer to the question of the most appropriate organizational structure for basic and applied statistical R&D. In this section, we briefly describe some possibilities for the organization of research in the Census Bureau; we offer these suggestions based in part on our reading of *Principles and Practices for a Federal Statistical Agency* (National Research Council, 2009b) and in part on the experiences of members of our panel in the management and oversight of censuses and complex survey operations.

### 4–C.1  Leadership

It is essential to have someone at the level of top management of a statistical agency who provides overall leadership for the technical side of the agency's work and who can articulate and defend the resources needed for basic research and applied methodology. This person should be responsible for methodology and statistical standards, as well as for informatics. It is extremely useful for this individual to be a noted expert in statistical methodology, who therefore can speak authoritatively about the importance of research and methodology not only in broad terms, but also in the context of particular projects.

At the Census Bureau, the appropriate level for this position is the associate director, a senior executive service position. Indeed, the top advocate of sound methodology in the Census Bureau until recently was the associate director for methodology and standards, a position that existed within the Bureau as far back as 1929 (under different names—see Box 4-1). But this position was abolished in 2005 in response to a refusal by the Department of Commerce to appoint the person recommended for the position by the Census Bureau director and deputy and, previously, was left vacant for several years after the resignation of the associate director.

### 4–C.2    Organization

The R&D function is organized in different ways in different national statistical organizations. In some, it is distributed to individual divisions responsible for a given program or subject, such as education or labor. In others, it is distributed to divisions with responsibility for broad subject-matter fields (e.g., demographic or business statistics). In still others, it is more fully centralized, reporting to the equivalent of an associate director of the Census Bureau and organizationally independent of subject-matter or field operations areas.

There are arguments in favor both of centralization and of decentralization. Decentralization can facilitate the integration of methodology into daily practice. However, since the operational entities are typically not headed by methodologists, this model tends to result in lower hierarchical positions for the heads of these decentralized methodology units, which makes it more difficult for them to assume a leadership function. Also, a lack of critical mass makes it more difficult to support specialization and basic research and to maintain high-quality standards for research and practice. Conversely, a centralized model is at greater risk of isolation from the daily practice of the agency, potentially endangering the viability of this function.

The Census Bureau seems at present to have the worst of both worlds. The Bureau's applied methodology work is decentralized, so there is no central leadership speaking on its behalf, yet its basic research is centralized and even more cut off from the rest of the Bureau than research tends to be intrinsically (see Box 4-1). The lack of central leadership for R&D at the top of the Bureau makes it difficult to integrate the work of the applied statisticians and the researchers with each other and with operational practice; it also makes it nearly impossible to plan research that supports fundamental, long-term changes.

Centralization has the following advantages that are useful to retain. First, it supports the professional independence and functional leadership of applied methodology. While methodologists need to be full and valued members of project teams (that is, staff groups who are working on method-

**Box 4-1** Historical Overview of the Census Bureau's Organization of R&D

**Early Years**

- *1902*—Permanent Census Bureau established
- *1909*—Census Bureau director is authorized to appoint a chief statistician, geographer, appointment clerk, private secretary, two stenographers, and eight expert chiefs of division, without examination by the then-Department of Commerce and Labor; the director and the assistant director remained presidential appointees with Senate confirmation (Magnuson, 2000a:136).
- *1929*—Secretary of Commerce is authorized to appoint two assistant directors, upon the recommendation of the Census Bureau director, one to serve as the executive assistant to the director and the other to serve as the technical and statistical advisor to the director—that person must have experience in statistical work (Magnuson, 2000b:139–140).
- *1933*—Census Bureau assistant director sets up the predecessor to the Statistical Research Division to achieve the goal of the Committee on Government Statistics and Information Services (COGSIS), formed by the American Statistical Association and the Social Science Research Council in early 1933, to create a research arm of the Census Bureau (Anderson, 1988).

**1950–1980 Censuses** (intercensal changes in names and responsibilities of directorates and divisions are omitted)

- *1950*—Assistant director for statistical standards, Morris Hansen, has responsibility "for statistical techniques throughout the Bureau. The personnel in this office worked in a staff capacity with the Assistant Directors and the divisions on many phases of the censuses. This office was responsible for the technical direction of the sampling, quality control, research and experimental work on methods and related activities; for developing and advising on publication practices and standards; and for the Post-Enumeration Survey, which was taken to evaluate the quality of the censuses" (U.S. Census Bureau, 1955:2). The 1950 census is coordinated by staff under the assistant director for demographic fields, Conrad Taeuber.
- *1960*—Assistant director for research and development, Morris Hansen, supervises the Statistical Research Division, Statistical Reports Division, and Electronic Systems Division. The Statistical Research Division "provided technical direction of the research, standards, and evaluation activities, and conducted research on the general census procedures during the 10-year interval between the 1950 and 1960 population censuses. Their work included research on and initial development of innovations in enumeration procedures and data-processing equipment and techniques as well as the sample design and other phases of the censuses" (U.S. Census Bureau, 1966:3). The 1960 census is coordinated by staff under the assistant director for demographic fields, Conrad Taeuber, who supervises the Population, Housing, and Agriculture Divisions, the Decennial Operations Division, and the Statistical Methods Division, which provided technical guidance on the long-form-sample design, quality control for data processing, and the 1960 research and evaluation program.
- *1970*—Associate director for statistical standards and methodology, Joseph Daly, supervises the Statistical Research Division and (as of 1971) a Research Center for Measurement Methods. Associate director for demographic fields, Conrad Taueber, supervises the Agriculture, Demographic Surveys, Foreign Demographic Analysis, Housing, Population, and Statistical Methods Divisions; the 1970 census is coordinated by staff under this directorate (U.S. Census Bureau, 1976:App. C).

*(continued)*

**Box 4-1**  (continued)

- *1980*—Associate director for statistical standards and methodology, Barbara Bailar, supervises the Statistical Research Division and Center for Survey Methods Research; the 1980 census is coordinated by staff under the Associate Director for Demographic Fields, George Hall and later William Butz.

    *Note*—From about 1960, the Statistical Methods Division staff are responsible for most basic and applied research on design, sampling, estimation, and other topics throughout the Demographic Directorate, reporting both to the Associate Director for Demographic Fields and the Associate Director for Statistical Standards and Methodology in a matrix management style. The Statistical Research Division is more directly involved in research for the Economic Directorate (personnel communication from Daniel Levine to the panel).

- *1987*—Associate director for statistical standards and methodology, Barbara Bailar, resigns in protest against the Department of Commerce's decision to abandon plans for a postenumeration survey that might permit adjustment of the 1990 census results for measured net undercount. The position is left vacant until 1990. About the same time, coordination for the 1990 census is moved out of the Demographic Directorate and into a new Decennial Census Directorate, headed by Charles Jones. This directorate establishes a Statistical Support Division, later the Decennial Statistical Studies Division; the reorganization reduces the influence of the Statistical Standards Directorate on census methodology.

- *1992*—Plannning, Research, and Evaluation Division (PRED) is established in the Statistical Standards and Methodology Directorate; PRED is focused on the decennial census and designs the 1998 dress rehearsal and 2000 census experiments and evaluations.

- *1994*—Center for Survey Methods Research (CMSR), which included behavioral and social scientists and focused on questionnaire design, measurement error, interviewer selection and training, and nonresponse, is abolished as a separate division within the Statistical Standards and Methodology Directorate, and the staff moved back into the Statistical Research Division.

- *2005*—Associate director for statistical standards and methodology position is left unfilled after the Department of Commerce fails to approve the Census Bureau's recommended candidate, so there is no senior management director in charge of R&D; the directorate's units are reassigned as follows:
    - Statistical Research Division (SRD) is retained and assigned to the deputy director;
    - Four senior scientist positions are assigned to the deputy director;
    - PRED is disbanded and the staff moved into other divisions, including Decennial Statistical Studies Division, Demographic Statistical Methods Division, and a new Data Integration Division for administrative records research in the Demographic Programs Directorate; and
    - Computer Assisted Survey Research Office (CASRO) is disbanded and the staff moved into the Technologies Management Office in the Information Technology Directorate.

**Current Organization**

- *Senior Scientists*—Report to deputy director; two of four positions are currently vacant

*(continued)*

---

**Box 4-1** (continued)

- *Statistical Research Division*—Reports to deputy director; staff of about 80 people, including about 12 students and postdocs and about 5 academics, all of whom have part-time appointments; includes eleven branches under three assistant division chiefs, one each for:
    - machine learning and computational statistics, computing applications, and missing data methods research;
    - sampling research, small area estimation research, disclosure avoidance research, and time series research; and
    - questionnaire design and measurement research, language and measurement research, questionnaire pretesting for household surveys, and human factors and usability research.
- *Decennial Statistical Studies Division*—Reports to assistant director for ACS and decennial census; staff of about 200 people; provides statistical support to the decennial census, including coverage measurement, and to the American Community Survey.
- *Demographic Statistical Methods Division*—Reports to associate director for demographic programs; staff of about 130 people; provides statistical support (sampling design, weighting, variance estimation, evaluation) to the portfolio of more than 30 household surveys conducted by the Census Bureau (e.g., Current Population Survey Annual Social and Economic Supplement, Survey of Income and Program Participation).
- *Economic Statistical Methods and Programming Division (ESMPD)*—Reports to assistant director for economic programs; staff of about 230 people; provides statistical support (sampling design, weighting, variance estimation, evaluation) to the Census Bureau's economic censuses and surveys (recently, several ESMPD staff were transferred to the Governments Division as part of a reorganization of that division's portfolio).

SOURCES: U.S. Census Bureau (1955, 1966, 1976, 1989, 1993, 1995a,b, 1996); for current staff counts, searches of the staff directory on the Census Bureau web site, http://www.census.gov, on January 20, 2010.

---

ological applications for components of specific programs, such as sample design and weighting for a particular survey), at the same time it is crucial that the methodologists receive expert guidance and technical supervision. This can best be achieved in a centralized organization in which the hierarchical position of everyone is strongly influenced by his or her technical competence. Professional independence is also vitally important, since on the rare occasion in which it makes a difference, these staff should be able to assert themselves and appeal, on professional grounds, decisions that are made within their project team with which they strongly disagree.

### 4–C.3 Project Teams

The contribution of methodology to an applied project, as well as the funds needed to finance the project, should be considered in the planning process before the project gets under way. This includes an assessment of the costs involving the contribution of all members of the team to the project, including methodologists, subject-matter specialists, and operations and IT people as appropriate, and a broad project plan that is formulated by high-level specialists from the participating disciplines.

On these projects, methodologists perform two general functions. First, at a strategic level, they help to ensure that the overall plan strikes an optimal balance between costs, timeliness, and respondent burden constraints on one hand, and other desired outcomes, especially improvements in data quality, on the other. While this is a leadership function and involves the entire project team, it is the methodologists who provide the framework and techniques enabling the team to grapple with trade-offs that must be considered. At a more tactical level, the methodologist is concerned with providing the statistical methods that are to be incorporated into the overall project design, which may include sample design, weighting, quality control, editing and imputation strategies, estimation, and analytic methods.

While the tactical contributions of methodology are easily understood, it is the strategic contributions that most benefit from leadership and sophistication and therefore support use of a centralized approach to manage R&D. Furthermore, in the case of major efforts, which will be directed by higher-level groups that may include directors of the participating divisions, a centralized approach will make it easier to judge major trade-offs and to resolve any conflicts with such an approach.

### 4–C.4   Funding

Funding of the basic research and applied methodology unit(s) should provide for pure research, applied methodological work, developmental projects, and maintenance work (quality control, routine reviews of edit failures, variance estimation, and minor design adjustments), together with supplementary resources from requests from operational units for additional methodological work. It is essential that there be a sound planning process that ensures that the funding needed to provide R&D support to the top priority basic and applied research projects for the agency as a whole and for particular programs, such as the decennial census, is obtained.

### 4–C.5   Training

Training must be a substantial portion of the budget for R&D, with additional emphasis on career development. This can be carried out not only in formal courses internally, but also through professional education courses at conferences, etc. Training serves a multitude of purposes. Most important, it should not only inculcate a basic knowledge of all that is involved, but also drive home the critical importance of teamwork and respect for the professional contributions of all the relevant disciplines.

### 4–C.6 Advisory Committees

A key tool for developing best practices and integrating them into the daily work of the organization is the effective use of advisory committees. Such committees can be used to provide critiques of all significant basic and applied research projects. Such critiques provide not only important contributions to the design and analysis of specific projects, but also a type of training for staff and validation based on the approval of the members given their professional standing. To be effective, advisory committees need to be given substantive information and important issues to address. In addition, their work needs to be buttressed by arrangements for bringing outside experts into the organization for intensive collaboration with in-house research staff.

### 4–C.7 Opportunities to Participate in Research

Most basic research should be conducted in partnership with applied methodologists to help ensure that the research carried out is relevant and that the results have the best opportunity to lead to changes in practice. Cooperative project work also helps with morale: while not everyone wants to do research (or is able to do so), a number of staff want to try their hand at it. And the very act of conducting some research, by those capable of it, leads to more open mind sets and a better informed practice.

## 4–D A NEW CENSUS RESEARCH AND DEVELOPMENT PROGRAM

### 4–D.1 Organization and Leadership

Consistent with the above discussion, the Census Bureau, as a high priority, should reorganize its basic research and applied methodology functions and how research and applied methods units interact with operational units. The objective should be to ensure that sound methodology pervades census and survey practice and to make sure that research programs are motivated by strategic issues facing the bureau. To inform an appropriate reorganization, the Census Bureau should undertake a fast-track, high-level management review of how research and development is organized in other national statistical offices and leading survey research organizations in academia and the private sector.

*Recommendation 4.1:* **The Census Bureau should comprehensively review the research and development practices and orga-**

nization in other national statistics offices and in survey organizations in academia and the private sector, with the goal of modernizing and strengthening the Bureau's own research and development program. Such a review should include assessments of and recommendations about:

- How to organize and direct basic and applied methods research to best serve the decennial census and other Census Bureau programs;
- How to organize information technology and database management to best serve research and operations, including how to manage the development of new technologies and ensure access to adequate expertise in these technical areas;
- How to operate collaborative project teams to facilitate timely innovation;
- How to ensure adequate training in survey methods and related fields;
- How to achieve extensive and intensive interaction with external research organizations and academic departments so that Census Bureau researchers and methodologists can benefit from related research work and ideas elsewhere; and
- How to fund and establish priorities for research and applied methodology work.

To carry out the findings of this review, the Census Bureau should consider reestablishing and filling an associate director–level executive staff position to head the statistical and survey research activities at the Census Bureau, with authority to organize the Bureau's research and applied methods activities. This position should have line authority for the basic research function. If the Census Bureau decides to adopt a centralized R&D model, it should also have line authority for the applied methodology function. If the Census Bureau decides to retain a decentralized structure for applied methodology work, the associate director position should have strong functional authority for the applied methods staff, including input on recruitment, promotion, and training of staff, quality standards, and project priorities. The position should have sufficient authority to ensure that research findings play a fundamental role in decisionmaking on the design of the decennial census and other major data collection programs. Given the scale and importance of the decennial census, this position should also have the authority for setting the census R&D agenda, which would include the selection of census experiments and evaluations.

*Recommendation 4.2:* To carry out the findings from the review recommended above, the Census Bureau should consider reestablishing and filling an associate director–level executive staff position to head the statistical and survey research activities at the Census Bureau, with authority to organize the Bureau's research and applied methods activities.

In addition, the Census Bureau should consider reestablishing the Center for Survey Methods Research as a unit under an associate director for statistical and survey research to conduct research on census and survey data collection instruments. This unit, which had a proud history of important research on questionnaire design, residency rules, and ethnographic research, no longer exists as a separate entity (see Box 4-1). Moreover, the subunits of the Statistical Research Division that engage in questionnaire design and measurement research, language and measurement research, questionnaire pretesting for household surveys, and human factors and usability research are no longer headed by a researcher of national reputation.

*Recommendation 4.3:* The Census Bureau should give greater emphasis to survey methodology. One possibility for doing so would be to establish a core survey methods research center, staffed by full-time survey researchers and headed by a nationally recognized expert in census and survey data collection instruments. Such a high-profile center could give priority to research on making effective initial contacts with census and survey respondents, including those made with new technologies.

In this report, we focus principally on methodology and operations research and not substantive analysis—basic research in social sciences. Nonetheless, we strongly support substantive research programs by statistical agencies in the subjects covered by their data collections. Such research is one of the best ways for an agency to obtain input on social, economic, and other kinds of changes that necessitate rethinking data collection and processing methods and the kinds of data that need to be provided to data users; basic research can be an important source of innovative ideas. For example, we echo the comments by the National Research Council (2006:175), recommending an office for research on population changes in geographic location and family living arrangements that relate to census residence rules and have implications for effective enumeration procedures. Substantive research by agency analysts should be relevant to policy and public information needs, although it should not take policy positions or be designed to focus on any particular policy agenda.

## 4–D.2   Integration

We have noted the importance that basic research be collaborative with applied methods research and that the latter be integrated with operations. We have also noted that research findings need to drive strategic decisions about census and survey operations. To achieve these goals requires that operational staff welcome and act on research results, which can be difficult when they are in the midst of data collection and processing and are under budget and timing constraints. The integration of research into the daily life of the Census Bureau should be the joint responsibility of the director and the associate director responsible for R&D, and it should be facilitated by a planning process that sets aside a block of funds for basic research, rendering explicit the unresolved development issues that need to be addressed for a given project to have a sound basis—and allocating the funds required. It is further incumbent on the leadership of a statistical agency to put in place incentives and structures so that research is integrated with operational planning. Such incentives might take the form of performance criteria and rewards for operational leaders who are assiduous in integrating research into their planning and, vice versa, for research leaders who are assiduous in remaining relevant to the operational needs of the agency.

In addition, the Census Bureau has an opportunity and an obligation to thoroughly integrate decennial census with ACS research. For the first time in census history, the ACS affords a continuous test bed not only for its own needs, but also for the decennial census, covering contact strategies, questionnaire design, data capture technology, and data processing. Although there are significant differences between the two programs, there are sufficient commonalities that basic and applied research and development needs to be conducted with continuous cross-fertilization between them.

> **Recommendation 4.4:** The Census Bureau should put in place incentives and structures so that research is fully integrated and collaborative not only across programs, but also with operational planning. Research should be responsive to operational needs, and, in turn, research findings should play a primary role in informing operational decision making.

> **Recommendation 4.5:** The Census Bureau should integrate decennial census and American Community Survey research—for example, by using the ACS methods panel as a test bed for the Internet and other data collection methods to consider in the census and by matching census and ACS records to evaluate coverage in both programs. To support comparative census-ACS research and to inform users, the Census Bureau should carry out analyses that explore, at both the aggregate level and the level of individual households, the degree of differences and the source

of differences in demographic characteristics and residence between the ACS and the decennial census.

### 4–D.3 Fostering Outside Collaboration

We have stressed, and cannot stress enough, the importance of extensive and intensive collaboration of in-house R&D staff with outside experts. No in-house R&D program can or should be sufficient unto itself. The attempt to do so is wasteful of scarce resources—whether the outcome is to reinvent the wheel or, even worse, to fail to make improvements in methods because of lack of familiarity with advances in other organizations, including leading survey and computer science research centers in academia and the private sector. In developing these relationships with advisory committees and external researchers, it is important that the Census Bureau view them less as a means of oversight and more as legitimate collaborators in the study and improvement of census operations.

> *Recommendation 4.6:* The Census Bureau should renew and augment mechanisms for obtaining external expertise from leading researchers and practitioners in survey and census methodology and in relevant computer science fields. These mechanisms might include (1) a more active census professional advisory committee program in which the members have an opportunity to work more closely with Census Bureau staff in developing and evaluating ideas for improved census and survey methods; (2) increased opportunities for sabbaticals at the Bureau for university faculty and other short-term appointments for both senior- and junior-level (graduate student) academics at the Census Bureau; (3) increased opportunities for sabbaticals for Census Bureau staff at academic institutions and private-sector survey organizations; (4) the awarding of design contracts early in the decade to support research and development of innovative technologies for census and survey data collection and processing; and (5) more effective use of contracting processes to obtain expert services.

### 4–D.4 Budgeting for Research

A complication for the Census Bureau's decennial census research program is the budget process. The timeline of the decennial census is such that it—and its level of spending, the extent of its coverage of and programs for specific population subgroups, and so forth—is a matter of intense attention in the time period immediately around the census year. However, that attention by a wide range of census stakeholders—including Congress, other executive branch agencies, and advocacy and interest groups—can drop off in the years following a census count. So too can the funds appropriated to the Census Bureau—a result that can restrict or preclude serious research and early planning for the next census.

The decennial census is necessarily a high-stakes program, and to some extent the escalating costs of the census and the steady accretion of coverage improvement operations (without a review of their cost-effectiveness) described in Chapter 2 result from this pressured environment. Absent the resources to conduct research on strategic design issues early in a decade—to guide the selection of principal design components and test the feasibility and interoperability of new and alternative methods—incrementalism in approach to the census is virtually inevitable. To their credit, Congress and presidential administrations have historically been unstinting in providing resources for the census as decennial dates have drawn close; the challenge going forward is to make the case that investment in research early in the decade—and the changes that develop from that research—will yield a more efficient and effective census in the end. Likewise, a Census Bureau research program should engage the entire range of stakeholders throughout the decade on key research and quality issues rather than try to pile on last-minute changes in years ending in 8 or 9.

Our urging in Recommendation 2.1 that the Census Bureau commit to bold and public cost and quality goals for the 2020 census is meant to promote a commitment to change early in the decade. We close this report on directions toward a new vision for the 2020 census by suggesting that national conversations on the nature of the census—and the research needed to effect real change—need to take place early, and over the whole decade.

> *Recommendation 4.7:* **The Census Bureau's planning for the 2020 census, particularly for research in the period 2010–2015, should be designed to permit proper evaluation of significant innovations and alternatives to the current decennial census design that will accomplish substantial cost savings in 2020 without impairing census quality. Otherwise, the census design in 2020 will either be an incremental change from that in 2010 with increased costs, or the Census Bureau may be compelled to implement a poorly evaluated and tested alternative design under severe time and cost constraints with a risk of substantially reduced quality. All involved, including Congress and the administration, should recognize that substantial cost savings in 2020 can be achieved only through effective planning over the course of the 2010–2020 decade and should fund and pursue research efforts commensurately.**

# – A –

## Past Census Research Programs

The descriptions of census research in this section generally exclude operations and analyses directly related to census coverage measurement—matching of the results of an independent postenumeration survey to census records in order to estimate undercount and overcount. We do, however, try to describe some of the formative work along these lines in the 1950 and 1960 censuses, given the novelty of the approach in those counts. Furthermore, although these descriptions do describe experiment and evaluation work related to census content that was part of the long-form-sample questionnaire prior to 2010, we exclude some work related to areas that are now fully out of scope of the decennial census (in particular, the census of agriculture).

### A–1  1950 CENSUS

#### A–1.a  Principal Pretests and Experiments Conducted Prior to the Census

#### Supplements to the Current Population Survey (CPS)

- *March 1946 (CPS areas, nationwide):* test of collection of both current residence (where the interviewer found a respondent at time of interview) and usual residence (the CPS and decennial census standard) information; provided particular information on enumeration of non-residents staying with households and college students.
- *April 1948 (CPS areas, nationwide):* tests of method of obtaining income data and of enumeration of people by both usual and current

residence. The test resulted in determination of pattern for asking income questions in 1950 as well as decisions (confirming residence rules) on enumeration of nonresidents and college students.

- *May 1948 (CPS areas, nationwide):* test of questions on physical characteristics of dwellings; led to revised definition of "dwelling unit."

### Experiments Conducted in Special Censuses or Other Surveys

- *April 1946 (Wilmington, NC):* experiment conducted as part of special census, focusing on collection of both current and usual residence information. Changes to enumerator training and questions on general population characteristics were also tested. The test resulted in first draft of population questions for 1950.
- *February 1948 (Washington, DC):* experiment conducted as part of survey conducted for National Park and Planning Commission, Bureau of Labor Statistics, and Housing and Home Finance Agency. The experiment focused on questions on income and led to revision in the format of the schedule, specific questions, and instructions.
- *May 1948 (Little Rock and North Little Rock, AR):* experiment conducted as part of special census on self-enumeration techniques; test resulted in information on possible response rates and comparative costs.
- *June 1948 (Philadelphia, PA):* experiment attached to survey conducted for Interdepartmental Subcommittee on Housing Adequacy on methods of measuring housing quality; test resulted in revised definition of "dilapidation."
- *March 1949 (Chicago, IL, and adjacent counties):* experiment conducted as part of Chicago Community Survey on rostering and obtaining complete enumeration of persons in households. Significantly, the experiment tested the use of a household questionnaire rather than either the master ledger-size schedule then used for census interviewing or individual person questionnaires.
- *June 1949 (Baltimore, MD):* test to check the quality of reported housing data conducted as part of survey conducted for Baltimore Housing Authority; test led to some revision of housing questions.

### Tests of Specific Phases or Operations of 1950 Censuses

- *May 1947 (Altoona, PA; Charlotte, NC; Cincinnati, OH; Louisville, KY):* test on "document sensing," or a format for the schedule to enable cards to be punched automatically; test indicated that technique was possible.

- *January 1948 (6 southern counties):* test of special Landlord-Tenant Operations Questionnaire led to revision of procedures.
- *August 1949 (33 field offices):* test of alternative population and housing schedules provided input to determination of final schedule (questionnaire) for 1950.
- *September 1949 (Puerto Rico):* test of population, housing, and agriculture questions for enumeration in Puerto Rico.
- *October 1949 (Raleigh and Roxboro, NC):* test of training procedures led to determination of final training plan.
- *November 1949 (Raleigh, NC):* test of questions on separate Survey of Residential Financing to be conducted simultaneously with the 1950 census.
- *January 1950 (Chicago, IL):* test of Survey of Residential Financing questions helped determine final procedures.

### Dress Rehearsals

- *April–May 1948 (Cape Girardeau and Perry Counties, MO):* test intended to (1) compare quality of data obtained from a schedule pared down to very few questions to one with many questions; (2) collect both current and usual residence information; and (3) generally assess quality of data from new questions. The pretest led to the conclusion that the short schedule yielded no substantial gains in quality over the longer, more-questions instrument. It also helped refine residence rules for some census types and specification of the duties of enumeration crew leaders.
- *October 1948 (Oldham and Carroll Counties, KY; Putnam and Union Counties, IL; Minneapolis, MN):* test focused on different enumeration procedures (self-enumeration, distribution of materials by post office, etc.) and their effects on data quality. Based on the test, the Census Bureau decided to use self-enumeration in the Census of Agriculture. The test also yielded cost, time, and quality data for the different approaches.
- *May 1949 (Anderson City, School District 17, and Edgefield County, SC; Atlanta, GA; rural areas near each of 64 CPS field offices):* test of training methods and final questionnaires and training procedures led to some modification in procedures (including procedures for shipping supplies to local offices) and determination of procedures for the postenumeration survey.

SOURCE: Adapted from U.S. Census Bureau (1955:Table B, p. 6), with additional information from Appendix E of that document.

### A–1.b   Research, Experimentation, and Evaluation Program

Although the 1950 census was the first to include a structured experimental and evaluation program (Goldfield and Pemberton, 2000a), details of the precise evaluations conducted in 1950 are not generally available. Only slightly more information about the shape of specific experiments is detailed in the Census Bureau's procedural history for the 1950 census. In a short section titled "Experimental Areas," that history notes (U.S. Census Bureau, 1955:5):

> A number of variations in the procedures for collecting data were introduced in ten District Offices. These variations made possible a comparison of procedures under actual census conditions. The experimental areas were located in Ohio and Michigan. In six of these districts, the alternative procedures involved the use of a household schedule (instead of a line schedule for a number of households), of the household as a sampling unit (instead of the person), and of self-enumeration (instead of direct enumeration). In four of the districts, assignments were made to enumerators in such manner that the variation in response could be studied in terms of enumerator differences.

Remarking on the self-enumeration portion of the 1950 experiments, the procedural history for the 1960 census clarifies that the test areas were in Columbus, OH, and Lansing, MI, and that part of the experiment requested that individual respondents complete and mail back (on Census Day) the questionnaires left with them by field enumerators (U.S. Census Bureau, 1966:292). The impact of the 1950 experiment on the later adoption of mailout-mailback methodology in the census is discussed by Bailar (2000).

### A–2   1960 CENSUS

#### A–2.a   Principal Pretests and Experiments Conducted Prior to the Census

#### Supplements to the Current Population Survey (CPS)

- *November 1958:* test of definitions for unit of enumeration.
- *April 1959:* test of farm definitions.

#### Experiments Conducted in Special Censuses or Other Surveys

- *March–April 1957 (Yonkers, NY):* major experiment conducted as part of special census. The experiment had two principal objectives: (1) obtaining data on the cost of a two-visit interview process (and effects on data quality) and (2) testing direct entry of interview information on machine-readable forms, either by enumerators or household respondents. In particular, a two-interview approach was tested in which— for roughly every fourth household—a limited amount of information was collected in a first interview and a long-form questionnaire was left for completion (and eventual pickup by an enumerator). The test demonstrated that enumerators could readily use the computer-readable schedules in the field but that respondents found them more difficult to follow. The two-visit approaches were found to be very expensive because of the difficulty of finding household members at home when enumerators attempted their visits. However, the idea of independent listing of dwelling units by crew leaders as well as enumerators was found to have potential value for improving coverage. The computer-readable forms generated by the test also helped with debugging the collection and tabulation routines. The Yonkers test also gave the Census Bureau the chance to test a new question on address of place of work, but the Bureau found the results to be unsatisfactory.

- *October 1957 (Indianapolis, IN):* focused test of several possible coverage improvement techniques in one postal zone as part of a special census. Building on the Yonkers test, one approach compared results of a recheck of listed addresses by crew leaders with enumerators' original results. Another technique involved preparing postcards based on enumerator interviews, with the postal carriers instructed to report boxes for which there were no cards for further verification (deliberately withholding a small sample of cards as a test of the carriers). Finally, the Bureau tested the completeness of lists of old-age assistance beneficiaries, juvenile delinquents, and other persons believed to live in low-income families obtained from local authorities, as well as the distribution of census forms in public schools for parents to complete. The tests provided evidence of coverage gains from the post office check and independent crew leader listings and suggested that the locally provided special lists were useful in indicating missed persons.

- *October 1957 (Philadelphia, PA):* census experiment focused on wording and placement of questions on labor force status, address of place of work, and date of marriage. Three different schedules were used in the experiment, which deliberately sent inexperienced interviewers to do initial questioning and trained CPS interviewers for verification interviews. One tested approach for the place of work question asked respondents to identify the location on a map of the city. The test resulted in revisions of the occupation and industry questions in the final census schedule.

- *November 1957 (Hartford City, IN):* test of mailout census methodology as part of a special census. A four-page Advance Census Report was mailed to individual household addresses; respondents were asked to complete the form but not to mail it back, holding it for an enumerator's visit instead. In the test, about 40 percent of households had completed part or all of the questionnaire prior to the visit, and the advance questionnaire seemed to be particularly useful in improving the quality of data on value of home.

- *January 1958 (Memphis, TN):* test of self-enumeration and mailback methodology as part of special census. Housing units were listed by enumerators and questionnaires distributed, with instructions to complete and mail on the census date. Mailed returns were chosen for follow-up verification, some by telephone and others by trained CPS interviewers; the post office check used in the Indianapolis test was also used for portions of Memphis. Part of the Memphis test also experimented with collection of information on visitors present in households on the census date and using reported usual residence information to try to allocate them to their usual home. Based on this test, a process for collecting "usual home elsewhere" information for some transient populations (enumerated where they are found on Census Day) was used in the 1960 census and in group quarters enumeration in subsequent censuses. Finally, the experiment asked some enumerators to query respondents for some information on neighboring living quarters (above or below, right or left, in back or in front); the technique was subsequently adopted as part of the evaluation process of the 1960 census.
- *February 1958 (Lynchburg, VA):* further testing of mailout of Advance Census Reports as part of special census. As in the Memphis test, CPS interviewers and crew leaders were used to reinterview some households to verify the quality of mailback data. Attempts were also made to time interviews to determine whether the Advance Census Report reduced the time spent in enumerator interviewing. In May 1958 enumerators returned to Lynchburg to recheck and verify the quality of housing data collected in February.
- *March 1958 (Dallas, TX):* census experiment on alternative questionnaire forms. Two small samples of households were interviewed using different schedules, specifically trying different methods for collecting information on income and place of work, on housing equipment items (e.g., type of heating and presence of air conditioning), and a 5-year versus 1-year migration question.
- *April, June 1958 (Ithaca, NY):* test conducted as part of special census (with staff follow-up) on enumerators' ability to classify types of living quarters.
- *October 1958 (Martinsburg, WV):* test intended to determine which of three alternative population and housing questionnaires could be used most effectively in the field. The content of the questionnaires was identical, and essentially identical to the questions that would be used in the 1960 census, but the alternative schedules varied the size and structure of the schedules. The Martinsburg test also delivered

Advance Census Reports prior to enumerator visits. Because the test was sufficiently close to the actual census, the test also permitted the Bureau to evaluate the training materials planned for use in 1960 as well as editing and coding routines.

- *December 1958 (Philadelphia, PA):* limited pretest to compare two possible forms with different skip patterns, after a final decision to use some self-enumeration in the 1960 census.
- *Mid-1959 (800 households in 10 regional headquarters cities):* small-scale test following the full-dress exercise in North Carolina (described below), testing variants of some questions to determine whether adding check boxes to routing questions (e.g., "If you do not live in a trailer, check here and continue with the next question.") promoted fuller response. Questionnaires were also distributed with return envelopes for mailback.

### "Informal" Experiments and Pretests for the Census of Housing

- *June 1957 (Washington, DC):* test of procedure for listing structures.
- *October 1957 (12 standard metropolitan areas):* test of collection of data on condition of housing unit in selected 1956 National Housing Inventory segments.
- *March 1958 (New York, NY):* test of classification of living quarters; used to formalize 1960 census definition of housing unit based on separate entrance or separate cooking equipment.
- *March, May 1958 (Prince George's County, MD):* tests of questions on exterior materials of housing and on basement shelters.
- *June 1958 (Lynchburg, VA):* recheck of data on condition and housing unit from May 1958 pretest.
- *September 1958 (Port Chester, NY):* recheck of data on classification of living quarters.

### Dress Rehearsal

- *February–March 1959 (Catawba and Rutherford Counties, NC):* final full-scale test prior to the census, making use of the Advance Census Report delivery that would be used in the 1960 count. Households chosen for the long-form sample were asked to complete their questionnaires and mail them to their local census office.

SOURCE: Adapted from U.S. Census Bureau (1966:App. B), with supplemental information from Part III, Chapters 1 and 2 of that document.

### A–2.b  Research, Experimentation, and Evaluation Program

A long list of possible research studies—based, in part, on input from the Panel of Statistical Consultants (an advisory group to the Census Bureau's assistant director for research and development)—for the 1960 census was refined to a final list of 22 studies. The Bureau divided these 22 studies into 8 "projects."

A. *Measurement of response variability*

- An expanded version of a study performed in 1950 to study variability in response due to enumerators and census staff; the 1960 version of the "Response Variance Study I" drew sample from all areas where questionnaires were mailed (with enumerators sent to pick them up or conduct interviews), rather than four selected areas as in 1950. The sample for which crew leaders and enumerators were assigned in order to estimate these effects included about 320,000 housing units and 1,000,000 persons.
- A follow-up study, called "Response Variance Study II," designed to measure variability in response due to the respondents themselves. A sample of 5,000 households from Response Variance Study I was drawn and enumerators sent to conduct interviews; a second sample of 1,000 housing units were asked to report again using a mailed, self-response questionnaire (with mailback, and interviewer follow-up if necessary).
- "Response Variance Study III" looked at variability due to coding and data entry. For a one-fourth sample of households for which data was collected in Response Variance Study I, photocopies of the enumeration books were made; pairs of coders then independently coded and transcribed the data.

B. *Reverse record checks (undercoverage in general population):* a check to see whether persons in an independent sample were enumerated in the 1960 census, in which the independent sample was culled from a mix of past census records and administrative data. Specifically, the sample was constructed from samples from four sources: (1) persons enumerated in the 1950 census; (2) immigrants and aliens registered in January 1960 with the Immigration and Naturalization Service; (3) birth records for children born between the 1950 and 1960 censuses; and (4) persons who were found in the 1950 postenumeration survey but not in the 1950 census.

C. *Reverse record checks II (undercoverage in specific groups):* a check similar to Project B, except that the records-based sample consisted of Social Security beneficiaries and students enrolled at colleges or universities.

D. *Reenumerative studies of coverage error*

- Postenumeration survey based on an area sample: a sample of "segments" that had been independently canvassed by enumerators for the Survey of Components of Change and Residential Finance (SCARF), conducted alongside the census. Provided with both 1960 census and SCARF information, enumerators in 2,500 area segments were tasked to search for omitted housing units or structures mistakenly labeled as housing units. About 10,000 housing units were administered a detailed housing questionnaire.
- Postenumeration survey based on a list sample: About 15,000 living quarters (both housing units and group quarters) already enumerated in the censuses was drawn, representing about 5,000 clusters of about 3 units each, dispersed across 2,400 enumeration districts. For each unit, a detailed reinterview was conducted in order to list persons within living quarters (this follow-up interview was conducted in early May 1960, so that not much time had passed since the April 1 Census Day). Enumerators were also asked to list "predecessor" and "successor" housing units or group quarters (e.g., neighboring units along a specified path of travel) in order to further check on missed housing units.

E. *Measurement of content error in data collection*

- Reinterview of about 5,000 households in the long-form sample for the 1960 census, using specially trained enumerators: a first phase (about 1,500 households in July 1960) sent the enumerators to conduct blind reinterviews; in the second phase (October 1960), dependent interviewing using the household's reported 1960 census information was conducted for half of the remaining sample and for the other half blind interviewing, as in the first phase.
- Reenumerative study focusing on housing characteristics, administering a battery of housing characteristics: about half of the 10,000 housing units in the study received the short form in the 1960 census, and the others were in the long-form sample (and hence had already reported some of the housing items).
- Match of records from the 1960 census long-form sample and the Current Population Survey's March or April 1960 samples.
- Match of respondent-provided occupation and industry information with data collected directly from employers.
- Match of about 10,000 sampled Internal Revenue Service returns to census records (although only about one-fourth of these were studied in depth regarding the consistency of reported income).

F. *Studies of processing error*

- A September 1963 detailed review of a sample of individually filled Advance Census Reports, Household Questionnaires (enumerator schedules), and the coded computer-readable forms in order to estimate transcription and other response errors.
- Coding error study in which a set of long-form questionnaires were separately coded by three coding clerks but only one was the designated "census coder" whose work went onto the final questionnaire.
- Study of the accuracy of automated editing rules in the microfilm–computer system for data collection and tabulation.

G. *Analytical studies:* general studies of census quality and coverage, including comparison of census counts with demographic analysis estimates.

H. *Post Office coverage improvement study:* postcensus postal check making use of postal employees and resources. A sample area of 10,000–15,000 housing units was selected in each of the 15 postal regions in the continental United States. These areas were matched to census enumeration districts. Postal carriers were asked to review name and address cards completed by enumerators for every counted household, making new cards for households on their routes that were not included in the census.

SOURCE: Adapted from U.S. Census Bureau (1963).

### A–3 1970 CENSUS

#### A–3.a Principal Pretests and Experiments Conducted Prior to the Census

#### Tests Conducted as Part of Special Census or Other Survey

- *August 1961 (Fort Smith, AR):* test conducted as part of special census to compare an address register compiled through enumerator visits with a register based on 1960 census records, new building permits, and a postal check. The separate listings were "found to be just as complete." Enumerator-visited households were also left with a brief questionnaire and asked to return the form by mail.

#### Pretests of Census Operations and Questionnaires

- *June 1962 (Fort Smith, AR, and Skokie, IL):* tests of address list updating using building permits and postal checks (updating the Fort Smith list from the August 1961 special census and repeating the methodology in Skokie). Short questionnaires were mailed to all households for response by mail, yielding 71–72 percent return rates.

- *April 1963 (Huntington, NY):* further testing of address listing and mailout-mailback methods, now focused on a larger city—deemed to be a rapidly growing area and including a mix of urban, suburban, and rural housing types. The test also included administering a long-form questionnaire to approximately 25 percent of households. In nonmail delivery areas, address lists were built through enumerator canvass and questionnaires deposited at the households for mailback.
- *May–June 1964 (Louisville, KY, Standard Metropolitan Statistical Area [SMSA]):* based on the Fort Smith, Skokie, and Huntington experiences, the Census Bureau requested and received funds for larger-scale testing of mailout-mailback methods (experimental censuses) in 1964 and 1965. The first of these was to be conducted in an area of about 750,000 population with a large central city; the Louisville area was selected (with the test being conducted from a special census office in Louisville, rather than the Bureau's nearby processing center in Jeffersonville, IN). The Louisville test added separate listing and contact strategies for "special places" (mainly group quarters), based on input from a task force on difficult-to-enumerate areas that had been convened in Louisville in late 1963. In this test, the Census Bureau concluded that a computer-generated address register based on previous census records and building permits was more complete than listing books completed by enumerators. In particular, "using a list based on enumerator canvass did not seem the ideal way to insure complete coverage in multiunit structures in city delivery areas." The Bureau further concluded "that at the current state of development of procedures the cost of a mail census, including certain important improvements over the 1960 approach, would not exceed the cost of a census by enumerator canvass."
- *April 1965 (Cleveland, OH):* Cleveland was chosen as the site for the second experimental mailout-mailback census both for its big-city nature and for its perceived enumeration problems in the 1960 census. For this test, the Census Bureau experimented with using a commercial mailing list as the base for the address register; the test was also the first to be geocoded by computer based on an address coding guide (i.e., coded information on address ranges on odd and even sides of street segments). As in Louisville, both short- and long-form questionnaires were distributed. The Cleveland test also centralized editing procedures, assigning completeness checks of returned questionnaires to district office clerks rather than individual enumerators. However, the Bureau's attempts to predesignate hard-to-enumerate areas were deemed to be less successful than was the case in Louisville, and probe questions added to the questionnaire on other households at the same street address were generally found to be unclear and confusing. Fur-

thermore, an attempt to test the effectiveness of rotating district office staff through different operations was also deemed unsuccessful, as many clerks balked at performing fieldwork as enumerators. Together, the Louisville and Cleveland tests convinced the Bureau that conducting at least part of the 1970 census by mailout-mailback was viable.

- *May 1966 (St. Louis Park, MN, and Yonkers, NY):* the first "content pretest" conducted principally by mail with little or no field follow-up. Questions on native tongue and national origin were of particular interest, and the two sites were chosen accordingly (predominantly Scandinavian origin in St. Louis Park and wide ethnic diversity in Yonkers). About 2,500 households in each area were drawn directly from 1960 census records to facilitate comparison of an "occupation six years ago" question with the reported occupation in the 1960 census. Editing and some follow-up rechecks were performed from district offices. The test questionnaires asked respondents for Social Security numbers, which were then compared with Social Security Administration records and generally found to be accurate.

- *May 1966 (national sample):* test of two alternative formats of a mail questionnaire (one presenting questions for each person on two facing pages and the other using a more traditional columnar format), administered to a national sample of about 2,300 housing units.

- *January–October 1966 (Wilmington, DE, SMSA):* further test to compare completeness of coverage of address registers compiled from commercial sources and postal checks with those formed by field canvassing. In different phases, the work in Wilmington focused on rural and locked-box addresses (comparing a January post office list with a field canvass in April–May) and city delivery addresses (having postal carriers check a commercial list in March, and sending enumerators to field verify "undeliverable" addresses in October).

- *January 1967 (Meigs, Morgan, Perry, Vinton, and Jackson Counties, OH):* test of alternative strategies for address canvassing, either knocking on every door or knocking only when necessary to verify the address or existence of housing units. The two approaches were eventually judged to have about equal effectiveness, but it was reasoned that this may be due to the preponderance of single-family structures (and that the knock-only-when-necessary approach might be problematic in areas with multiunit structures).

- *March 1967 (Memphis, TN):* test comparing intensive "blitz" enumeration by teams of enumerators with normal enumeration and interviewing procedures in a 14-tract (about 25,000 population) area characterized by low-income and deteriorating housing.

- *March 1967 Content Pretest (Gretna, LA):* test of employment questions (revised for compliance with new federal government standards) in two census tracts, chosen for having a high unemployment rate and above-average proportion of substandard housing. New items on mobility, marital history, birth dates of children ever born, and disability were also included on test questionnaires. The Census Bureau also received a list of persons arrested during the study period from the Gretna Police Department in an attempt to see if they could be matched to the pretest records, but such matches were generally unsuccessful.

- *April 1967 (New Haven, CT, SMSA):* further test of mailout-mailback census techniques, building on the Louisville and Cleveland tests, particularly focused on centralized office operations (as in Cleveland) and adding coverage improvement operations to field follow-up work. New questionnaire content, including Social Security number and vocational training, was also included on the test questionnaires. The qualifier "ever married" was dropped from the question on children ever born, which yielded surprisingly little reaction; however, the Bureau concluded that "New Haven respondents objected to the number of questions on bathroom facilities in particular, and the large number of sample questions in general." The address coding guide technique for geocoding results was further refined for the New Haven test, and a commercial mailing list (with prelisting or canvassing of rural delivery addresses) was the base for the address register. Among the coverage improvement techniques tried in this test were soliciting lists of people likely to be missed from community organizations and a "movers check" focused primarily on people who moved during the months before and after Census Day.

- *May 1967 Questionnaire Format Test (national sample):* test of four different designs of mailing pieces (e.g., foldout sheets versus booklets, variations in question formatting) administered to 4,900 urban housing units in a national sample. The version of the questionnaire modeled on that used in the New Haven experiment was judged to be most practical for both respondents and the Census Bureau.

- *August 1967 (Detroit, MI):* canvass of about 800 addresses (on 450 postal routes) in buildings with at least two housing units, intended to get a sense of variation in apartment numbering styles or, if unnumbered, what kind of locational labels might be used to describe individual units.

- *September 1967 (North Philadelphia, PA):* mailout exercise (using short- and long-form questionnaires) in two inner city tracts (about 21,000 population in total) where population was suspected to have dropped significantly. The test also involved recruiting temporary enumerators familiar with the area to collect questionnaires and look for missed buildings and living quarters. In particular, an attempt was made to use high school students as enumerators (with school counselor supervision), but this was found to involve numerous administrative problems (and encountered some resistance by respondents).

- *October 1967 (Kalamazoo, MI, SMSA):* focused study of address list development in "fringe" areas where city delivery and rural service were mixed. About 16,500 such housing units were found in the Kalamazoo area in this test, with about 20 percent being addresses that could be found on city-delivery listings and in the enumerator canvasses.
- *May 1968 Housing Quality Study (Austin, TX; Cleveland, OH; San Francisco, CA):* 300 housing units in each of three cities were interviewed and separately rated by American Public Health Association inspectors in order to assess a revised definition of "substandard housing."
- *August 1968 Subject Response Study (CPS areas, nationwide):* specific test of an "occupation 5 years ago" question, reinterviewing about 2,800 households from a 1963 CPS sample.

### Dress Rehearsals

- *May 1968 (Dane County, WI):* dress rehearsal of all major census processes using a decentralized mail census system, under which receipt, edit, control, and follow-up of questionnaires returned by mail would be distributed among enumerators in local offices.
- *May 1968 (Sumter and Chesterfield Counties, WI):* dress rehearsal of all major census operations using traditional nonmail techniques (Advance Census Reports were mailed to households, followed by enumerator visits to pick up questionnaires or conduct interviews).
- *September 1968 (Trenton, NJ):* dress rehearsal of all major census processes using a centralized mail census system, under which receipt, edit, and control of questionnaires returned by mail would be managed by clerks in census district offices. Furthermore, enumerators would visit only those households for which follow-up could not be carried out by telephone. A fourth dress rehearsal focused on hard-to-enumerate areas was originally planned for March 1969, but budget constraints forced some of these plans to be folded in with the Trenton dress rehearsal.

SOURCE: Adapted from U.S. Census Bureau (1976:Chap. 2).

### A–3.b   Research, Experimentation, and Evaluation Program

The 1970 Census Evaluation and Research Program consisted of 25 individual projects, falling under five broad headings:

- *Evaluation of coverage of persons and housing units*
    - Demographic analysis: comparison of census totals with population estimates based on previous census, birth and death registration, and estimates of migration, as done in 1950 and 1960.
    - Medicare record check: conducted in support of the demographic analysis results, which depend in part on Medicare registration rolls as an auxiliary data source for older age groups. A sample of about 8,500 people age 65 or over was drawn from Social Security Administration records and matched to 1970 census returns to estimate missed persons.
    - Birth registration study: a second project carried out in direct support of demographic analysis estimation. A sample of about 15,000 children born between 1964 and 1968 was derived from household interview records from the Current Population Survey and from the weekly Health Interview Survey conducted for the U.S. Department of Health, Education, and Welfare by the Census Bureau. The compiled vital statistics records of the National Center for Health Statistics—themselves drawn from birth and death certificate data from state and local registration areas— were then searched to try to find the sample children, with field follow-up visits if no birth record data could be found.
    - Housing unit coverage in mail areas: two-part study of the completeness of the address register used for the 1970 census mailing. In the first phase, permanent Census Bureau survey interviewers were tasked to inspect and visit a sample of 20,000 addresses from the mailing list to determine the number of living quarters at each address. In the second, survey enumerators were asked to relist about 8,000 city blocks (or similar-size equivalents in noncity areas).
    - Study of housing unit occupancy status and of census deletes: Interviews and information from the housing unit coverage study described above were also analyzed to try to sort out reasons for housing units erroneously marked as vacant, and the extent to which errors were caused by households maintaining two places of residence or moving during the census period.
    - Current Population Survey–Census match: match of CPS records collected during the week of March 19, 1970, to census returns in order to estimate the gross number of missed housing units at the overall level (not just the mailout areas, as in the housing unit coverage study). (The matched CPS–census records were also used to evaluate the quality of consistency of common data items, under the "measurement of content error" heading below.)

- Definitional errors in housing unit count: this study was intended to measure gross and net definitional error, defined as instances in which the occupants of multiple housing units were combined and counted as one household or the occupants of a single housing unit were split and counted as multiple households. Samples of about 140,000 questionnaires from mailout-mailback areas and 70,000 from enumerator-collected areas were clerically reviewed to identify households with potential for such errors, and field reinterviews scheduled with a subsample of those flagged cases.
- Special procedures for hard-to-enumerate areas: review and assessment of five specific coverage improvement programs implemented for the 1970 census: (1) movers operation, which asked post offices for 20 large cities for change-of-address information for people who moved a month before and after Census Day; (2) precanvass, a final precensus review of census mailing lists for the inner-city areas of the largest metropolitan areas; (3) "missed person" forms distributed to community action groups, asking them to identify persons that they believed may have been missed by the census; (4) postenumeration postal check of address listings in traditional enumeration areas (i.e., not mailout-mailback); and (5) vacancy recheck, a follow-up study of a nationwide sample of 15,000 housing units classified as vacant by enumerators.
- Analysis of census coverage by local residents: ethnographic study of census experiences and perceived reasons for undercounting in a sample of living quarters that was then being studied (along similar lines) by other Census Bureau demographic survey programs.
- District of Columbia driver's license study: records check of a sample of about 1,000 young males (ages 20–29) who had either newly obtained or renewed an existing driver's license in the District of Columbia between July 1969 and June 1970.

- *Measurement of content error*

  - National Edit Sample: a sample of about 15,000 households, distributed across the different types of enumeration areas (mailout-mailback and conventional enumeration). For mailout questionnaire packages in the sample, a special envelope was included so that the questionnaire was mailed directly to the Census Bureau's main processing center in Jeffersonville, IN; a photocopy of the questionnaire was made before sending it to the appropriate local office, and a copy of the final questionnaire after editing and follow-up steps was obtained for comparison with the mailed original. During the 1970 census, the National Edit Sample returns—having been funneled directly to Jeffersonville—were used as part of an "early warning" system for judging workload in local and district offices; after the census, the sample was analyzed to study the effects of editing operations.

- Sample control in mail areas: a study of how the final composition of the roughly 1-in-5 long-form sample compared with the original design of the sample (i.e., whether long- and short-form questionnaires were interchanged in multiunit structures).

- Quality control of field operations: review of quality control records maintained by both enumerator crew leaders and district office personnel to identify factors in cases in which enumerators or other staff had to be released due to poor work.

- Geographic coding evaluation: study in which enumerators were sent to verify the block, tract, and place codes automatically assigned by computer (using the Bureau's new electronic address coding guides) to a sample of about 5,000 census listings in mail areas. A second phase of the study sent a second enumerator to verify the geographic coding obtained by the original census interviewer in nonmail, traditional enumeration areas for about 5,000 listings.

- Quality of census sampling: review of counts in sample (long-form) and nonsample households by such factors as household size, race, sex, age, and household type.

- Coding quality: comparison of questionnaire items coded (on to microfilm/computer readable forms) by an independent coder.

- Place-of-work data: follow-up of the accuracy of reported place-of-work information (the 1970 census requested a full address of the workplace, whereas the 1960 census was less specific) for a sample of about 4,000 persons drawn from matched CPS–census records (described above).

- Content Reinterview Study: similar to reinterview studies in the 1950 and 1960 censuses, but asking detailed population and housing items of all reinterviewed households. A sample of about 11,000 housing units (10,000 occupied and 1,000 flagged as vacant) from the long-form sample universe was flagged for reinterview. Special attention in the reinterview analysis was paid to new questionnaire items on the 1970 census, such as mother tongue and vocational training. The content reinterview sample was also compared with vital statistics data to assess the accuracy of the last child born and number of children ever born items, and matches were made before and after editing, processing, and imputation procedures in order to study their impact.

- Disability study: specific evaluation study requested by the Social Security Administration, reinterviewing about 15,000 households in which at least one person was reported as disabled and about 25,000 households said to contain only nondisabled persons. The reinterview questionnaire contained additional probes and alternative wordings to try to elicit fuller response to the disability questions.
- Employer record check: similar to the check of reported occupation and industry data against employer records conducted in 1960 but with a larger sample size: about 6,000 persons rather than 2,000.
- Employment 5 years ago: report of the 1968 Subject Response Study (see Section A–3.a), in which data on employment, occupation, and industry reported by a sample of Current Population Survey interviewees in July 1963 were compared with answers to "occupation 5 years ago" questions on a 1968 test questionnaire.
- Record check on value of home: revival of a 1950 census reinterview study, wherein the actual sale price for about 3,000 single family homes sold between July–December 1971 was compared with the reported "value of home" question on the long-form sample questionnaire.
- Record check on gross rent: similar to the record check on home value, the study of gross rent in 1970 extended an idea from the 1950 census. In five selected metropolitan areas, about 1,200 rental-occupied households were drawn from the 1970 census returns; the local gas and electric utility companies provided data on amounts paid for each of the 12 months before Census Day, which were compared with census-reported figures.
- *Assessment of interviewer's contribution to census errors:* estimation of response variance due to enumerators based on a sample of enumeration assignments distributed across "decentralized mail" areas (which, combined, accounted for about one-half of the U.S. population; in the 1970 census, questionnaires in these areas were mailed to local or district offices, where enumerators edited and checked for completeness the returned questionnaires).
- *Assessment of publicity campaign:* a sample survey administered to about 600 radio stations and 700 television stations, asking about the extent to which their programming publicized census topics and the number of times that public service spots about the census were broadcast (including the dollar value of those spots, if they had been treated as paid commercials).

- *Experiment on expanded mailout-mailback enumeration:* anticipating that future censuses would expand mailout-mailback coverage to more than the densely populated urban areas targeted in 1970, a sample of 10 district offices in rural areas—paired so that one would be counted by traditional procedures (control) and the other attempted by mailout-mailback (test). Returns from these pairs of offices were then compared by coverage completeness, cost of enumeration, and item nonresponse.

SOURCE: Adapted from U.S. Census Bureau (1976:Chap. 14).


## A–4   1980 CENSUS

### A–4.a   Principal Pretests and Experiments Conducted Prior to the Census

### Tests Conducted as Part of Special Census or Other Survey

- *April 1975 (San Bernardino County, CA):* first test of use of computer terminals in local (district) offices for transmission of cost and progress reports and payroll information. The terminals were also used to send block-level population and housing unit counts, in part to facilitate a first-ever review of those counts (specifically, precensus estimates) by local authorities. Although some communication problems were found, similar computer configurations were used in later tests in Pima County, Travis County, Camden, and Oakland (described below). However, the Bureau ultimately decided not to place terminals in the local district offices during the 1980 census due to cost and maintenance issues. The local review of population and housing units was also tested in subsequent experiments, and, in the end, local authorities were given a chance to review counts between the first and second phases of follow-up during the 1980 census.
- *October 1975 (Pima County, AZ):* further testing of district office computer terminals and local review of preliminary housing unit and population counts, conducted as part of a special census. The Pima County test also focused on the use of "nonhousehold source" lists—lists of person names collected from local authorities. A list of about 2,700 names and addresses of predominantly Hispanic persons was collected from community organizations and from public assistance programs, and this list was cross-checked with the test census address register. About 6 percent (160) of listed addresses were found to have been missed by the test census; follow-up with these missed persons found an additional 231 persons who had been missed in the census count.

## Pretests of Census Operations and Questionnaires

- *April 1975 (Salem County, NJ):* test of the collection of income data on a 100 percent basis (that is, on the short form rather than the long form of the census). The test grew from dissatisfaction with the long-form sample data on income from the 1970 census, particularly for small areas due to wider implementation of revenue-sharing legislation. The Salem County test tried four different versions of an income question, including one simple "total income" question (with categorical response) as well as one modeled on the detailed ledger-type query usually found on the long-form questionnaire.

- *May 1975 (national sample):* national mail-only test of the four income questions tested on a local level in Salem County, NJ; the national test included 19,700 housing units, and some nonresponse follow-up was conducted by Census Bureau survey interviewers. The Salem County and national tests suggested the feasibility of including an income question on the short form—particularly a question combining a series of yes/no questions on income types with a categorical "total income" response—but the Bureau decided in 1977 to exclude the question from the 1980 census short form.

- *September 1975 (four counties in Arkansas, two parishes in Louisiana, and three counties in Mississippi):* test of three alternative strategies for address prelisting in rural areas: (P1) inquire only when necessary, knocking on doors to gather address information only if observation was unclear; (P2) inquire at every structure, with only one follow-up attempt if information could not be obtained from observation or from neighbors; and (P3) inquire at every structure, with "unlimited" (several) follow-up attempts, using information from neighbors only as a last resort. However, the P3 strategy was only "simulated" by making additional visits to P2 addresses where no homeowner was at home during the initial visit but information had already been determined by observation or from neighbors. The Bureau concluded that P2 should be used in the 1980 census (and it was), finding that it outperformed P1 but that additional gains from further callbacks in P3 were outweighed by additional costs. In a sample of enumeration districts in these areas, a separate quality control crew listed 25 addresses in each district and matched those to address registers collected during the test; a similar quality control measure would be used in the 1980 count.

- *Fall 1975–Winter 1976 (Columbus, OH):* test of the quality of address lists obtainable in Tape Address Register (TAR) areas, or mail delivery areas in urban centers where mailing lists could be obtained from commercially produced computer tapes. Such registers had been used in 145 metropolitan statistical areas in 1970 and were proposed for use in all available areas in 1980. The test compared attempts to update the 1970 census address register for Columbus with four different commercial TARs (as well as post office checks and Census Bureau geocoding). The Census Bureau found the TARs to be usable for 1980 but ultimately decided against trying to directly update 1970 files based on the new sources.

- *April 1976 (Travis County, TX):* first major pretest of census procedures, running from the opening of a district office in Austin in January through closing of the office in mid-September (about 2 months behind schedule). The Travis County test was intended as a "mini-census" involving both mail and field components. In the test, Census Bureau officials paid particular attention to monitoring the mail return rate to the local census office as a diagnostic tool. The questionnaires in the test census followed the 1970 census closely, save for revised questions on income (following up on the Salem County and national mail income pretests) and Hispanic origin. The Travis County test also began the innovation of providing Spanish-language questionnaires upon request; a Spanish-language message in the main mailing package directed interested respondents to call a telephone number or check a box on the English language form in order to request the Spanish questionnaire. However, take-up on the Spanish questionnaire was very low—only 50 requests out of 15,000 households with a Hispanic-origin householder. The Travis County test also replicated the "nonhousehold source" name and address list approach from the Pima County special census, comparing census coverage with names and addresses from driver's license files. The test also directly followed up with persons who had given a change-of-address notice to the post office within one month of Census Day.

- *May 1976 (Gallia and Meigs Counties, OH):* test of revised methods for designating enumeration districts (essentially, an individual enumerator's workload area) to ensure that they comply with natural, recognizable land features rather than invisible political boundaries. The test found little difference in results between using the visually recognizable enumeration districts and the block groups carved out of those districts for tabulation purposes. Hence, block groups were used as standard units in 1980.

- *July 1976 (national sample):* National Content Test involving mailed questionnaires to about 28,000 housing units divided into two panels. Each panel was administered a questionnaire containing alternative versions of several questions, including relationship, ethnic origin, education, school attendance, place of birth (asking about mother's place of residence at birth rather than actual birth location), and disability. About 2,300 households from each panel were chosen for a content reinterview by a Census Bureau survey interviewer in September–October 1976. The disability question, in particular, was found to fare poorly in accuracy as measured by the reinterview, but it was nonetheless retained in the 1980 census due to its policy importance.

- *September 1976 (Camden, NJ):* second major "minicensus" test, which focused heavily on new coverage improvement techniques for hard-to-enumerate areas. These techniques included (1) team enumeration, comparing the effectiveness of enumerators working alone, enumerators working in pairs, or blitz-type enumeration by a crew or team (the team methods were found to produce better quality results at the cost of some loss in productivity; in 1980, managers would be authorized to use team methods to resolve hard-to-enumerate areas); (2) Spanish-language forms available upon request, as in Travis County, but with expanded use of walk-in questionnaire assistance centers rather than phone banks; (3) "nonhousehold source" lists from driver's license rolls, as in Travis County; (4) formation of a complete-count committee of local officials and organizers to promote the census, modeled after a group that had formed in Detroit during the 1970 census; (5) fielding of a short survey to assess awareness of the advertising and public information campaign associated with the test census; and (6) testing of a two-phase local review process to examine preliminary housing unit and population counts.

  The Camden test is perhaps most notable because the city of Camden challenged the pretest results (and the Bureau's 1975 population estimate) for the city as being erroneously low, occasioning hearings before the U.S. House Subcommittee on Census and Population. The city also filed suit against the Census Bureau, seeking an injunction on use of the sub-100,000 population estimates for the city in federal and state allocation programs. Ultimately, the suit was dismissed in March 1980 by mutual consent of the city and the Census Bureau.

- *September 1976 (three chapters of the Navajo Indian Reservation, Arizona and New Mexico):* test of revised procedures for tribal lands, beginning with an enumeration of the three chapters by usual census methods. The results were then compared with the Navajo population register maintained by the Bureau of Indian Affairs (BIA) of the U.S. Department of the Interior; nonmatched cases were submitted for field follow-up. The BIA register was determined to be inadequate for use as a coverage improvement device in the 1980 census. However, the Census Bureau found BIA maps—combined with aerial photography—useful in mapping the areas and assigning enumerator work.

- *January 1977 (four counties in Arkansas, two parishes in Louisiana, and three counties in Mississippi):* "rural relist test" in which the same counties where address listing activities were conducted in September 1975 were revisited. The test was intended to provide information to choose between "early" address listing in the 1980 census (listing in spring 1979, followed by two postal checks) and "late" listing (listing in January 1980, with one postal check in March 1980). Late listing was found to have generally better coverage, but the two postal checks in the early listing scheme compensated for the coverage differences.

- *April 1977 (Oakland, CA):* third major pretest of major census operations, this one incorporating alternative questionnaire designs for the race and Hispanic-origin questions (in response to an Office of Management and Budget directive that data be provided for four race categories). The relationship question was also revised to ask about each person's relationship to a reference person (the person completing the form as "Person 1") rather than to the "head of household." In the test, the Census Bureau experimented with the use of reminder postcards to urge households to return their form; such cards were mailed to even-numbered enumeration districts. Although the test results implied that the reminder postcards could boost mail response by as much as 5 percent, the Bureau judged that the gains would not outweigh the additional postage cost and declined to use reminder postcards in the 1980 census. Operationally, the test also paid some enumerators using hourly rates rather than "piece rates," as had been done in 1970.
- *July 1978 (national sample):* based on responses to the Hispanic-origin question in the Richmond dress rehearsal (below), the Census Bureau conducted a National Test of Spanish Origin of about 3,200 housing units, by mail, during summer 1978. The test compared different versions of that question, and the question that was judged to have the best performance was slated to be used in the September 1978 lower Manhattan dress rehearsal (below).

## Dress Rehearsals

- *April 1978 (Richmond city and Chesterfield and Henrico Counties, VA):* on the recommendation of the Civil Service Commission (later the Office of Personnel Management), the Richmond dress rehearsal approached enumerator training from a "performance-oriented" standpoint, emphasizing visual aids and workbooks rather than verbatim recitation. The Richmond rehearsal was also the first test of a dependent household roster check (later used in the 1980 census) in which households returning incomplete questionnaires were recontacted by phone or enumerator visit; the roster of people listed on the original questionnaire was read back to the respondent to determine its accuracy. The questionnaire used in the Richmond rehearsal was also the first census questionnaire to use color printing to enhance readability (blue backgrounds to highlight questions and blue type for cover information and instructions).

- *April 1978 (La Plata and Montezuma Counties, CO; portions of Archuleta County, CO, and San Juan County, NM, were also included to complete coverage of the Ute Mountain and Southern Ute Indian reservations):* dress rehearsal featuring the first test of traditional door-to-door enumeration since the 1970 census. Specifically, Advance Census Reports were mailed to households, but respondents were instructed to keep them until an enumerator came to collect them. Because of the Indian reservations included in the dress rehearsal area, a supplementary questionnaire for on-reservation households was added to the test (and later used in the 1980 census). The supplementary questionnaire included questions on tribal affiliation, migration, utilization of government programs, and income. In the test, it was noted that reservation households were displeased if they had to complete both the census long-form questionnaire and the supplementary questions, so the supplement was waived for on-reservation long-form households in 1980.

- *September 1978 (Lower Manhattan, NY):* a third dress rehearsal was not part of the Bureau's original plans, but advisory committees challenged the racial and ethnic diversity of the Richmond and Colorado rehearsal areas (particularly for Hispanic and Asian American populations). In this test, the Census Bureau experienced resistance from mail carriers when a postal check of address lists was performed; the carriers rebelled at having to complete a separate postcard for each unit in large multiunit structures that had been omitted from the census address register (as a result of these criticisms, future postal checks permitted only one "add" postcard to be completed for all the units of a completely missed structure/address). Difficulties in recruiting temporary enumerators prompted the Census Bureau to seek a waiver of the requirement that census workers be U.S. citizens (a waiver that would later hold for the 1980 census).

SOURCE: Adapted from U.S. Census Bureau (1989:Chap. 2).

### A–4.b   Research, Experimentation, and Evaluation Program

For the 1980 and subsequent censuses, the descriptions of research and experimentation programs that follow do not include those studies related directly to coverage evaluation through a postenumeration survey. In all, the Census Bureau's program of evaluations and experiments included about 40 separate projects.

- *Coverage evaluation*
    - Housing unit coverage studies: program to estimate missed housing units based on a match of Current Population Survey and census returns as well as a follow-up interview with a sample of households already counted in the census (i.e., a subsample of the "E" or enumeration sample to which the postenumeration survey is compared). The studies also estimated the rate at which whole housing units containing at least one household member duplicated elsewhere were themselves duplicated in the census (i.e., through clerical or geographic coding error).
    - CPS–Census retrospective study: reverse record check study in which one rotation panel from the March 1977 CPS (about 20,000 people) was matched to the 1980 census returns, with the intent of determining how often people could not be either directly matched or successfully contacted (traced) to verify their address on Census Day 1980. Census clerks were permitted to examine 1979 Internal Revenue Service return data to try to find a different address if a sample person could not be found at their 1977 address. The study was conducted between 1982 and August 1983 (potentially putting further distance between people and their 1977 or 1980 addresses).
    - CPS–Internal Revenue Service (IRS) administrative records match: as part of initial research into the feasibility of triple-system estimation, about 92,000 records from the February 1978 CPS were matched to an IRS data file extract based on Social Security number (SSN). Records were also sent to the Social Security Administration for matching to summary earnings records, either by SSN (for those CPS records that included SSN) or based on name and other information, and match rates across the three data sources were estimated and compared.
    - IRS–Census direct match: the CPS–IRS records match suggested a surprisingly high erroneous match rate: cases matched based on SSN were found not to match on name. This result prompted a further study in which a sample of census records was matched to the IRS individual master file.

- *Coverage improvement evaluations* (in rough chronological order of the operations they describe; the growth in the number of these operations between 1970 and 1980 is discussed in Chapter 2, and the individual operations are further described in National Research Council, 1985):
    - Advance post office check: evaluation of a complete review by the U.S. Postal Service of a compiled set of commercial address

lists (about 38 million addresses). The Postal Service suggested adding 5 million addresses and deleting or modifying 2.9 million others. A match of the Postal Service–suggested additions to the Census Bureau's internal address register yielded a net of 2.2 million additions. (A small sample of about 4,100 addresses from the commercial lists was deliberately withheld, for later matching to the Postal Service's returned list; about two-thirds of the withheld addresses had been added by the postal check.) The operation cost about $7 million, about $5 million of which was paid to the Postal Service.

– Casing and time-of-delivery post office checks: evaluation of problems (e.g., missing addresses or known-undeliverable addresses) detected by local postal staff as they "cased" (sorted in delivery order) the census questionnaire mailing packages. Further reports were obtained from the local mail carriers as they attempted delivery. Together, these two checks resulted in the identification and enumeration of another 2 million housing units; the two operations cost about $9.3 million, about $6 million of which was paid to the Postal Service.

– Precanvass: evaluation of an operation in which enumerators canvassed areas carrying excerpts from the census address register (updated from the advance post office check) to verify or correct entries and add additional found housing units. The precanvass was conducted on a 100 percent basis in the urban areas where the commercial address list–based register was to be used; after the census, precanvass registers were compared with final census registers and census data for a sample of enumeration districts in order to study the characteristics of added housing units. The Census Bureau estimated that the precanvass added 2.36 million addresses to the census, at a cost of about $12 million.

– Casual count: evaluation of coverage improvement operation through which teams of two enumerators were sent to places that people with no permanent place of residence were expected to frequent: transit stations, unemployment offices, bars, and so forth. The enumerators attempted interviews to collect information and determine whether an interviewee might have been counted elsewhere. Ultimately, the procedure was found to have added only about 13,000 people (at a cost of about $250,000).

– Census questionnaire coverage items and dependent roster checks: field operation to resolve disparities in reported counts: i.e., mismatches between the reported number of persons in the household and the number of names listed in response to Question 1 on the questionnaire or between the reported and expected (from the address register) number of living quarters at the address. In the follow-up interviews, respondents were allowed and asked to review the names on the originally reported roster to make additions or deletions. The operation was conducted in 260 enumeration districts (systematic 1-in-1,000 sample of districts). The Census Bureau estimated that the living quarters check—comparing the reported number of living quarters at the address to the address register and following up on discrepancies—ultimately added about 93,000 housing units to the census, concentrated in 30 of the 260 sample enumeration districts.

– Whole household "usual home elsewhere" identification: the 1980 census questionnaire included a question after the main household roster asking whether everyone listed in the roster was only staying at the Census Day location temporarily and had a usual home elsewhere. Home addresses could be entered on the back cover of the questionnaire. The information was transcribed onto a new form and routed through the appropriate district office for the "usual home" location. Subsequently, the Bureau mounted a clerical check operation to verify whether these "usual home elsewhere" cases had in fact been counted at the usual place of residence. About 547,000 people in 301,000 housing units fell into the whole household "usual home elsewhere" category; the clerical check and further evaluation suggested that a total of about 1 million people were reallocated through this operation and about 200,000 were counted in at least two places (because their listings at temporary addresses had not been deleted).

– Nonhousehold sources program: similar to operations in previous tests, the Census Bureau compiled independent lists of persons (particularly minority populations) that sources believed might be missed in the census. In particular, the 1980 version of the program used driver's license records, U.S. Immigration and Naturalization Service records, and 1979 New York City public assistance files as sources of possible missed persons. The actual yield of persons added to the census—about 1.9 percent of the compiled lists—fell well short of the 10 percent that pretests had suggested; however, about 58,000 persons were eventually determined to be cases that should have been added to the census but were not due to processing time constraints. The public assistance and Immigration and Naturalization Service files were judged to yield about twice as many adds per follow-up case as the motor vehicle files.

– Follow-up of vacant and deleted housing units: evaluation of procedure for visiting every housing unit flagged as "vacant" or "deleted (as nonexistent)" by census enumerators. About 5.5 million vacant and 2.3 million deleted units were covered by this operation; about 10 percent and 7.5 percent of the vacant and deleted units, respectively, were found to be classified incorrectly and converted to occupied units. At $36.3 million, the vacant and delete check was the most costly of the coverage improvement programs in the 1980 census.

– Prelist recanvass: evaluation of the completeness of the address list in prelist areas (those for which the address list was principally generated and verified prior to the census by enumerator visit). In 137 district offices, areas were selected to be recanvassed and listed; these recanvass registers were then compared with the census master address register to determine whether any housing units could be added or deleted. In the sampled areas, about 105,000 housing units containing about 217,000 persons were added to the census due to the recanvass; the operation cost $10.29 million.

– Assistance centers: assessment of the level of usage of walk-in questionnaire assistance centers in 87 centralized district offices and telephone assistance for all 373 district offices. Evaluation was incomplete because necessary records were not retained; however, the evaluation suggested that about 790,000 contacts had been made for questionnaire assistance, most commonly on residence questions (who to list on the roster) and income reporting.

– Spanish-language questionnaire: evaluation of the usage of the Spanish-language questionnaire that could be requested in mailout-mailback areas (either by a check box with Spanish instructions on the main English-language questionnaire or through contact with a census assistance center or nonresponse follow-up interviewer) or from the enumerator in nonmail areas. Necessary data for this evaluation were incomplete: enumerators were not asked to keep a record of the number of Spanish-language questionnaires they distributed on their rounds.

– "Were You Counted?" program: evaluation of a program through which people who believed themselves to be missed by the census could fill a blank questionnaire and return it to the Census Bureau. A blank "Were You Counted?" questionnaire was given to urban newspapers (including some with questions translated into various foreign languages) for publication as a public service; the form could be clipped and mailed for processing.

About 62,000 forms were received nationwide, reporting data for about 140,000 people; about one-quarter of these were found to have been already enumerated and about half were added to the census (the remaining quarter were unusable, including those that provided an unlocatable address).

- Postenumeration post office check: evaluation of a postcensus check in which local Postal Service staff were asked to review the addresses in which conventional door-to-door, list-enumerate methodology had been used. This review suggested another 148,000 housing units that might have been missed by the census; ultimately, the Census Bureau estimated that this operation added about 50,000 housing units and 130,000 people to the final census counts.

- Local review: evaluation of a procedure in which preliminary population and housing unit counts were generated down to the enumeration district level, after completion of nonresponse follow-up, and given to local officials to review. The officials were asked to identify and provide evidence of any major discrepancies between the counts and their local records and estimates. About one-third of 39,000 contacted governmental units opted to participate in local review, and about one-half of those flagged at least one discrepancy. About 20,000 discrepant cases were able to be resolved without resort to recanvass in the field; recanvass work corrected geographic coding for about 28,000 housing units and added about 53,000 housing units to the census.

• *Content evaluation*

- Content reinterview study: reinterview of about 14,000 housing units that had been included in the long-form sample, using a questionnaire focused on questions that were new or substantially revised for the 1980 census (e.g., school of attendance, non-English language spoken and ability to speak English, ancestry). Additional probe questions or alternative question language was added for several of the items. Interviews were conducted by members of the Census Bureau's CPS staff (permanent interviewers) between November 1980 and January 1981; proxy reporting was allowed for young people (age 15 or less), with proxy interviewers for persons over age 16 allowed only as a last resort. About 88 percent of the sample was successfully reinterviewed. Questions were evaluated for their response variability and for unusual patterns that might suggest reporting problems (i.e., finding that Idaho was commonly reported as the place of birth in the census when the reinterview indicated either Illinois or Indiana, suggesting a coding or handwriting interpretation problem).

– Analysis of matched census–content reinterview study records: two separate analysis during the 1980s of the data (including results of alternative question wordings) from the content reinterview study. A 1983 study focused on education-related questions (e.g., highest grade attained, degrees attained, present enrollment); a 1986 study examined responses to the place-of-birth question in more detail.

– Utility cost follow-up surveys: comparison of reported data on utility costs in the 1980 census (asking about the previous 12 months) with actual cost data obtained from 11 utility companies in 8 cities in December 1979. The participating companies sent a notice of each customer's average monthly utility costs to half of their residential customers along with their March 1980 bills (thus providing them with an accurate answer to the census question); the other half of the customers served as a control group. The companies then submitted lists of customer names, addresses, average monthly utility cost, and sample/control group status to the Census Bureau for matching to census returns. The data from the experiment were analyzed in 1982, leading to the finding of general overreporting of utility costs (more so for gas than electricity); the advance notification (sample group) reduced the size of the overreporting but did not eliminate it.

• *Processing and quality control evaluations:*

– Evaluation of qualification tests for coders: prospective employees for coding operations (matching respondents' reported information for industry and occupation or for place-of-work/migration) were put through multistage qualification tests, using decks of census questionnaires filled with artificial data. The evaluation studied performance at the different stages (i.e., improvement in accuracy on a second, more difficult deck of test cases conditional on results of the first deck).

– Coding accuracy: quality control checks were performed for coding of items onto machine-readable forms, differentiating between the particularly detailed coding for industry and occupation and place-of-work/migration and "general" coding of all other items (e.g., ancestry, income). For income-item coding, a special analysis considered the frequency of "factor of 10" errors: inadvertent shifting of a decimal point to the left or right.

– Geographic Base File/Dual Independent Map Encoding (GBF/DIME) file closeout evaluation: to review the accuracy of the GBF/DIME files used to assign geographic identifiers (e.g., ZIP code, census tract, and city/place) to addresses, a stratified sample of 600–800 addresses from each of the 280 GBF/DIME files was drawn. Cards were generated for each sample address and sent to the appropriate regional census offices; the geographic codes in census files were compared with the GBF/DIME results by a geographic specialist in each regional office.

- *Other studies*
  - Census Logistical Early Warning Sample (CLEWS): similar to the National Edit Sample of the 1970 census research program (see Section A–3.b), the CLEWS was a national sample of 6,000 households (evenly divided between short-form and long-form-sample households) for which the mail response envelope was specially marked ("CLEWS") and addressed for direct return to the Census Bureau's processing center in Jeffersonville, IN. A copy of the questionnaire was made and retained in Jeffersonville before forwarding to the appropriate district office; the CLEWS was used to provide quick estimates of daily mail-return rates and detect possible enumeration problems. The CLEWS questionnaires were also used as the control group for the alternative questionnaire experiment described below.
  - Imputation, allocation, and substitution: Census Bureau staff developed evaluation studies for the three basic methods for assigning data for unreported items (for housing, persons, or both). In support of the evaluation, census field staff undertook a verification check of about 11,000 units in 12 areas with high rates of "unclassified" housing units (those lacking information on household size or clear indication of vacancy status). Part of the evaluation also considered the extent to which the data used to fill gaps for particular housing units wound up being drawn from an adjacent unit—that is, the housing unit immediately preceding the "gap" unit on the Census Bureau's record tapes.
  - Public information evaluation:
    * Advertising media evaluations: audits commissioned by the Advertising Council to estimate the reach and frequency of the advertising materials (distributed to media outlets for dissemination on a nonpaid, public-service basis) for the 1980 census. In particular, the auditors were asked to estimate the value of the broadcast or publication of the advertising materials if they had been run on a paid basis. Separate studies were arranged to study the dissemination of the materials among predominantly black and Hispanic audiences.

* Knowledge, attitudes, and practices survey: special surveys prior to the census (late January–early February 1980) and after the census (late March) asking six questions about the census, the confidentiality of census data, and the uses of census data for apportionment and redistricting. The study was intended to assess the degree to which the public-service advertising materials increased awareness of basic census concepts. The precensus survey reached about 2,431 housing units (out of 3,772 eligible for the sample), and the postcensus survey included returns from 2,446 interviews (out of 3,115 eligible).

– Applied Behavior Analysis Study: experiment in which permanent Census Bureau survey enumerators were sent to visit a sample of about 11,000 housing units (clustered in 20 district offices) in the short window between Census Day and the start of nonresponse follow-up activities. Respondents were asked to describe how they had or were participating in the census (e.g., acknowledging receiving the form in the mail, starting to fill in the form, mailing in the form). The study was intended to study factors differentiating those households that had already mailed in the form from those who merely acknowledged receiving it. Later, the interview records were matched to final census records to study whether people who reported not receiving the form had in fact been counted and whether households reporting that they had mailed the form were actually counted in the census on an enumerator return (nonresponse follow-up).

• *Experiments*

– Alternative questionnaire experiment: test of two sets (both short and long form) of questionnaires. One set was designed to be machine readable and presented the short-form questions in horizontal rows rather than vertical columns; long-form content was slightly rearranged. The second set was not machine readable by the Bureau's then-current processing system but emphasized visual appeal and respondent comprehension. Although the experimental groups and control group were intended to include about 18,000 addresses, only 14,400 returns were usable due to packaging and delivery problems.

– Telephone nonresponse follow-up experiment: experiment to test the feasibility of nonresponse follow-up by telephone in mail census areas and to compare the cost of telephone and personal visit methods. The study was limited to single-unit structures because the "crisscross" telephone directories used to obtain telephone numbers could not provide apartment or subunit designations; moreover, the study did not involve northeast or south census regions because crisscross directories were not available. From the workloads of selected district offices, a total of 1,000 nonresponse cases were processed in each of four treatment groups: short-form telephone, short-form personal visit, long-form telephone, and long-form personal visit.

– Update List-Leave experiment: test of an alternative type of enumeration, conducted simultaneously with the main census effort between March 11 and March 26, 1980. Enumerators were tasked to canvass an enumeration district, simultaneously comparing and updating address register entries and leaving a short- or long-form questionnaire for respondents to complete and mail back. Any group quarters encountered during these canvasses were reported to the enumeration crew leader and handled separately. Five district offices (Dayton, OH; northeast central Chicago, IL; Yakima, WA; Greenville, NC; and Abilene, TX) were chosen for the experiment and paired with control offices in nearby areas.

– Studies of employee selection
  * Selection aids validation study: program of testing, both as part of the Oakland and lower Manhattan census tests and through in-house study, of selection procedures for temporary census employees. In particular, the tests administered to prospective temporary census employees (nonsupervisory and supervisory) were analyzed for fairness to racial, ethnic, and sex groups.
  * Adverse-impact determination study: review of the disposition of about 62,000 applicants from a sample of district offices in order to determine whether employment rates by demographic subgroup were consistent with civil rights law (i.e., that no subgroup faced an "adverse impact" as defined in the Civil Rights Act of 1964, as amended).
  * Predictive validity project: further assessment of the test administered to prospective 1980 census enumerators, this time to study the test's predictive power for enumerator longevity (number of weeks on the job and completion of task) and productivity (e.g., time spent per acceptably completed census form).

– Evaluation of training methods
  * Experimental student internship program: feasibility study of using college students as census enumerators in return for academic credit (as determined by the school) and pay

during their 6-week appointment. A total of 46 campuses participated in the experiment; in all, about 1,500 students, faculty members, and Census Bureau staff were involved in the test; about 30 participants traveled to Washington, DC, for a postcensus evaluation workshop in November 1980. Although the interns tended to complete their assignments at a higher rate than the general enumerator staff, the Census Bureau concluded that they were not as productive as the enumerators and were not available for the expected 30-hours-per-week detail.

* Alternative training experiment: comparison of the default 1980 census training routine (verbatim reading of a training guide by a designated trainer) with alternative training materials (i.e., checklists, flow charts) modeled after those used in industry and the armed forces. Three matched pairs of district offices were chosen for the experiment; in all, about 1,400 received training in the control group (verbatim) and 1,200 in the experimental group (alternative method). The alternative techniques (particularly group learning activities such as role playing) were ultimately found to result in better job performance.

* Job-enrichment training experiment: feasibility study conducted in one of three district offices in Dallas, TX, in which about 70 out of 150 newly hired temporary enumerators accepted an offer to represent the Census Bureau at local community meetings. Those who did so (and, hence, publicly declared their Census Bureau affiliation) were more likely to stay on the job through completion of nonresponse follow-up than those who did not, controlling for other factors.

SOURCE: Adapted from U.S. Census Bureau (1989:Chap. 9).

## A–5  1990 CENSUS

### A–5.a  Principal Pretests and Experiments Conducted Prior to the Census

### Pretests of Census Operations and Questionnaires

- *July–August 1983 (Essex County, MA):* prior to the widespread availability of geographic positioning systems (GPS) technology, the Census Bureau conducted tests using the U.S. Coast Guard's Long Range Navigational System (LORAN-C) to record the geographic location of rural residences during the 1990 census. Like GPS, LORAN-C used low frequency radio signals to derive geographic locations; however, it had the key limitation that its coverage was largely limited to coastal areas in the continental United States. In the early 1980s, the question was whether it was feasible to use a LORAN-C receiver unit in the field—particularly when, at the time, the estimated cost of a census-ready LORAN-C receiver was as high as $1,000 per unit. Early research by the Census Bureau in 1983 suggested that a custom-built solution would be necessary because existing off-the-shelf LORAN-C receivers did not meet the Bureau's specifications. Tests in Massachusetts in summer 1983 examined the ability to obtain coordinates from a car-based receiver. Based on poor results, the Census Bureau began to examine the then-developing GPS network, but the full set of GPS satellites was not expected to be in place until 1987–1989. In May 1984, the Bureau decided to suspend further direct research on navigation systems in the 1990 census.

- *1984 (Bridgeport and Hartford, CT; Hardin County, TX; and Gordon and Murray Counties, GA):* multiple-site Address List Compilation Test in both urban (Connecticut) and rural (Texas and Georgia) areas to assess the quality of various address list development technologies. In the Connecticut portion of the test, three basic address list sources were used and compared—commercial vendor mailing lists, U.S. Postal Service lists, and the 1980 census address register updated through new construction permits and other sources. In Hartford, all three lists were obtained and subjected to a "precanvass" updating by Census Bureau field staff; in Bridgeport, the Postal Service lists were not used, and the commercial and 1980-updated lists were put through a mail carrier check and the census field precanvass. In the rural areas, the basic strategies compared were (1) a complete prelist canvass by Census Bureau field staff updated and corrected through a postal carrier check and (2) a Postal Service–generated list updated by census field staff. Based on the test, the Census Bureau decided to use commercial vendor lists for urban areas (using a postal carrier check and a census precanvass operation) and a complete prelisting canvass in rural areas (with a postal check).

- *March 1985 (Jersey City, NJ, and Tampa, FL):* first major operational test, intended to focus primarily on increased data collection automation that had been developed since the 1980 census and on a two-stage collection strategy in hard-to-enumerate areas. In the Tampa test, the primary focus was on the use of optical mark recognition in data processing (with the computer-scanned results being compared with traditional keying of questionnaires in Jersey City); it also focused on other attempts at adding automation to census processes and the effectiveness of a reminder postcard. The Jersey City test emphasized a two-stage collection method that separated out the collection and nonresponse follow-up operations for the short-form and long-form data. In the first stage, short forms were mailed to all addresses (with an enumerator visit to nonresponding households), and long-form questionnaires were sent to a 1-in-5 sample in the second stage (with enumerator follow-up). (In all but a small sample, the short-form questions were asked again in the long-form interviews; in the sample, the only short-form item collected was respondent's name.) The goal was to see whether this operational switch would have benefit in traditionally hard-to-count areas. However, in June 1985, the Census Bureau opted to cancel the nonresponse follow-up for the long-form, second stage collection due to extremely low mail response rates in the two-stage test areas (and, accordingly, a greater nonresponse follow-up workload involving more costly long-form interviews).

- *June 1985 (Chicago, IL):* informal, special test of race and Hispanic-origin questions as a preview for the National Content Test in 1986 (see below). The test considered two short-form designs (one of them the 1980 census model as a control). The principal variation in the questions was the addition of a general "Asian or Pacific Islander" category and a "Yes, Spanish/Hispanic" choice—permitting write-in of specific ethnic origins, such as Mexican or Honduran—to the Hispanic-origin question. Enumerator follow-up was conducted on a limited basis, using experienced Census Bureau survey interviewers; the intent of this follow-up was less to produce estimates than to gather personal observations from the interviewers about respondents' processing of the questions.

- *April 1986 (national sample):* National Content Test of 46,000 housing units testing new question wording and formatting, particularly for the race and Hispanic-origin questions. The alternative questionnaires in this test also included a multiple residence coverage probe question ("Does this person regularly live at another residence for 30 or more days during the year?") and short-form housing questions including year built, acreage and use, value, and year built. The test also varied the mailing package itself—one envelope designed to be attractive to the eye and the other meant to appear more official. Mailout of test questionnaires was delayed by two weeks because the initial shape and thickness of the mailing packages caused automated assembly hardware to malfunction (hence requiring questionnaires to be inserted by hand into envelopes). Field nonresponse follow-up was conducted on a 25 percent sample basis; a further subsample of these cases was contacted yet again to evaluate consistency of responses. Based on the test, the Bureau decided to move questions on marital history and age at first marriage from the short form to the long form, also adding military service, disability, and some housing items to the long form.

- *March 1986 (central Los Angeles County, CA; eight counties in east central Mississippi, including the Choctaw Indian Reservation):* second major operational pretest of census procedures. In Los Angeles, the test attempted to provide information on the use of a local processing office separate from the local census (data collection) office and the effectiveness of the Census Bureau's planned coverage measurement and statistical adjustment procedures. In Mississippi, the major foci of the test were address list development and questionnaire delivery techniques for rural areas and specific enumeration procedures for American Indian reservations. In particular, the Mississippi test included the first use of update-list-leave enumeration in which enumerators visited areas and simultaneously updated address listings (and map locations) and delivered questionnaires. Favorable results led to the method being repeated in the 1988 dress rehearsal (and the 1990 census). The Mississippi test also included the Census Bureau's first use of laptop computers for address listing; despite good results, the Census Bureau decided against using portable computers in the 1990 census due to the cost of the equipment. In late March, the Census Bureau canceled census test activities in the Compton office, one of the two district data collection offices in the Los Angeles test area, due to abnormally low mail response.
- *April 1987 (north central North Dakota, including the Fort Totten and Turtle Mountain Indian Reservations):* third major test census, in an area of about 75,000 population. In addition to continuing to refine enumeration procedures for Indian reservations, the test also sought to determine whether mailout-mailback methods could be used in "prelist pockets" of mail delivery areas embedded in areas that the Bureau would typically count through door-to-door enumerator visits. The test also sought to make the enumerator schedule/questionnaire attractive and easy to follow (whereas previous questionnaire design had been focused on the main mailout questionnaire). The North Dakota test also continued to vary and test different strategies for enumerator pay, including bonus pay for enumerators and crew leaders.
- *June 1987 (six metropolitan statistical areas—Los Angeles–Long Beach, San Francisco–Oakland, and San Diego, CA; Houston, TX; New York, NY; Miami, FL):* further testing of race and Hispanic-origin questions, conducted by mailout to about 27,000 housing units in six metropolitan areas known to contain significant concentrations of Asian or Pacific Islander, Cuban, Mexican, or Puerto Rican populations. This mailout test was dubbed the Special Urban Survey.
- *September–October 1987 (Honolulu, HI; Anchorage and Bethel, AK; El Paso and San Antonio, TX; Charleston, WV):* following the Special Urban Survey, the Census Bureau convened focus group interviews

in six cities across the country to assess comprehension of the race and Hispanic-origin questions for selected demographic groups. In Honolulu, separate focus groups of Asians and Pacific Islanders were held; the Alaska focus groups included Eskimos, Aleuts, and Alaska Natives; the Texas focus groups focused on Hispanic populations; and the West Virginia site included a mix of blacks and whites. Each focus group included 8–10 participants.

- *October 1989 (targeted sample):* special survey conducted as a result of concern among the Asian and Pacific Islander communities, and interest from Congress, in improving the quality of data for those groups. Based on this pressure, the Census Bureau decided to add specific examples of Asian and Pacific Islander categories to the 1990 census form and to permit write-in entries for specific categories; the special survey was a test of this last-minute change. The survey was administered to a sample of about 40,000 housing units mainly targeted to areas with high levels of American Indian and Asian and Pacific Islander response in 1980 (Philadelphia, Chicago, Detroit, New York, and San Diego) as well as a sample from rural areas in West Virginia and Mississippi.

## Dress Rehearsal

- *March 1988 (St. Louis, MO; 14 counties in east central Missouri; 8 counties in eastern Washington, including the Colville and Spokane Indian Reservations):* dress rehearsals in all three sites provided the first opportunity to use the Census Bureau's new TIGER database to produce maps and for enumerators to provide TIGER updates. The rehearsal also featured revised training procedures for operations requiring door-to-door visits, emphasizing role-playing simulations in addition to verbatim recitation. The S-Night operation tested in the dress rehearsal was a first, multiple-step procedure to try to include segments of the homeless population in the census count.

  The rehearsal was also the first full-scale systems test for the computers and systems developed for use in district offices, processing centers, and headquarters; delays and bid protests in the solicitation for development of these systems meant that the contract had not been awarded until May 1987. In testing during the rehearsal, census staff found problems with the generation of automated management information reports as well as sluggish response times in the local census office computers when multiple operations were performed simultaneously. Concerns that the systems might not have the memory needed for full-scale use in the 1990 census led to a crash software development program to resolve problems. The automation of census processing did enable the Bureau to test a "search/match" coverage improvement operation, trying to match persons reported on some census forms (group quarters residents indicating a "usual home elsewhere," military census reports filled by armed forces personnel, and blank "Were You Counted?" forms available from public places) to a usual home of residence.

In the east central Missouri site, nine counties that were originally intended to be enumerated through mailout-mailback yielded large numbers of undeliverable addresses in an advance postal check of the address register; these were converted to update-list-leave areas where enumerators visited households (updating address entries as necessary) and left questionnaires for mailback. The St. Louis portion of the rehearsal introduced a variant of update-list-leave, called urban update-leave, in hard-to-enumerate urban areas, testing the procedure's use in public housing areas of the city.

SOURCE: Adapted from U.S. Census Bureau (1993:Chap. 2).

### A–5.b Research, Experimentation, and Evaluation Program

Formally known as the 1990 Census Research, Evaluation, and Experimental (REX) program, the Census Bureau's slate of evaluations and experiments centered around 17 studies (including work specific to the Post-Enumeration Survey and possible adjustment of census figures for nonresponse). Modifications to and extensions of the original 17 studies were proposed and approved by a cross-division steering committee as late as August 1990.

- *Data content and quality*
  - Alternative Questionnaires Experiment: test of five different questionnaires (each administered to about 7,000 households) varying question wording and sequence; the alternative questionnaires also varied as to whether they included a "motivational insert" to explain the purpose of the census and try to boost mail return rates. The experiment included a control group of 6,000 households that received the standard census questionnaire.
  - Master Trace Study: a study planned as one of the original 17 REX programs but one that was ultimately abandoned "due to operational difficulties and budget constraints" (U.S. Census Bureau, 1996:11-15). As recommended by the National Research Council (1988), the Census Bureau planned to trace a national sample of 31,000 questionnaires through all stages of operations. In addition to computer files, physical copies of forms and documents were to be made and retained before and after operations that were not computer automated. By design, all households

chosen for the content reinterview study (described below) were to be included in the Master Trace Study sample.

- Content reinterview study: reinterview of about 13,000 households in September–December 1990, to assess response variability of selected census questions. The reinterviews were conducted principally by telephone rather than field enumerator visit, with sample households alerted of the impending call by a letter.
- Census–Residential Finance Survey match: evaluation of consistency of responses to relevant census items (i.e., income) among households visited by the Census Bureau's separate Residential Finance Survey.
- Macro-level consistency check: a planned study that was eventually abandoned due to budgetary and time constraints, the macro-level consistency check was intended to compare 1990 census data (disaggregated by demographic characteristics) with non–Census Bureau sources. These external sources may have included the Current Population Survey as well as administrative data from states and from the federal Medicare program.
- Coding evaluation: quality assurance operation making use of more experienced coding personnel to ensure that selected write-in responses were keyed and categorized correctly.
- Integrated evaluation of error: evaluation study intended to develop a "total error model" for 1990 census data, breaking down census error by demographic characteristics and documenting the contributions of various census operations to overall error.

- *Coverage* (not including studies directly related to the 1990 Post-Enumeration Survey)

    - Ethnographic evaluation of behavioral causes of undercount: as recommended by the National Research Council (1978, 1985), the Census Bureau commissioned an extensive suite of observational studies by outside researchers, chronicling responses and attitudes toward the census by numerous traditionally hard-to-enumerate populations. Joint statistical agreement contracts were executed with researchers to study 29 selected areas in the continental United States and Puerto Rico. The ethnographers were also asked to conduct their own "alternative enumerations" of their sample areas (using their own methodology) for comparison with the official census counts. Areas (and demographic populations) covered by the ethnographic studies included inner-city areas with high crime rates or major migration shifts, as well as studies of populations on American Indian reservations and migrant agricultural workers in farm communities.

- Assessments of coverage improvement measures: analysis of evaluation data for each of more than 20 address list building or coverage improvement operations. Studies requiring additional data collection or that focused on new or significantly revised operations include:
  * Address list development: evaluation measures for address list development followed the 1980 census model closely. A systematic sample of addresses was withheld from the address list submitted to the U.S. Postal Service in the initial Advance Post Office Check operation; Census Bureau field staff were also dispatched to visit a sample of addresses flagged by the local post offices or individual mail carriers in time-of-delivery postal checks. The 1990 program allowed local officials to review block-level housing unit counts both prior to and following the main 1990 count. Precensus local review was opened only to jurisdictions in mailout-mailback areas (with about 16 percent of eligible jurisdictions opting to participate and 84 percent of those challenging their precensus block counts). About 25 percent of all functioning governmental units participated in the postcensal local review; about 67 percent of those issued challenges. The precensus local review was judged to be particularly effective, with recanvass of challenged blocks ultimately yielding 367,313 additional housing units to the census roster.
  * Urban update-leave: following the procedure's use in the St. Louis portion of the 1988 dress rehearsal (see Section A–5.a), urban update-leave was conducted in 346 blocks in Chicago, Detroit, Los Angeles, Baltimore, Cleveland, and Philadelphia. In the operation, enumerators were asked to simultaneously verify (or correct) the address register and leave census questionnaires for later mail return. The operation was intended to focus on blocks comprised principally of large public housing developments; however, evaluation suggested that a higher-than-expected 20.7 percent of the units covered in urban update-leave areas were single unit structures.
  * Urban Update-Enumerate: similar to urban update-leave, but focusing on entire census blocks preidentified as containing mainly boarded-up units; the intent was to avoid including those units and structures in the larger vacant/delete check. Conducted in only 96 blocks in New York City and Detroit, the operation proved difficult because regional census offices had difficulty identifying blocks meeting the "boarded up" criterion (e.g., identifying blocks that were expected to be

fully vacant due to large construction projects but that turned out to be occupied due to construction delays).

* Shelter and Street Night (S-Night): to evaluate the intensive S-Night operation intended to cover parts of the homeless population as well as users of emergency and temporary assistance shelters, external researchers placed teams of 60 observers in areas expected to be reached by street enumerators. S-Night was conducted on March 20–21, 1990; the evaluation teams were placed in samples of sites in Chicago, Los Angeles, New Orleans, New York City (with 120 observers), and Phoenix. A separate piece of the evaluation studied the quality of the list of shelters that was developed to conduct S-Night operations.

* Vacant/Delete/Movers Check: the vacant/delete check procedure used in the 1980 census (see Section A–4.b) was expanded to try to identify (and correctly assign geographic locations) to people who moved residences on or shortly after Census Day.

* Primary selection algorithm review: the Census Bureau's primary selection algorithm is an automated routine for choosing the "best" questionnaire in instances in which data are obtained on multiple forms for units with the same identification number. The review operation scrutinized the questionnaires that were *not* identified as the "best" form in order to find any people who might have been missed in the final count. The clerical review ultimately yielded 350,448 people to the census, on a review of 401,174 multiple-questionnaire cases.

* Search/Match: expanding the whole-household "usual home elsewhere" search program from the 1980 census (see Section A–4.b), the 1990 incarnation tried to reconcile "usual home elsewhere" address information reported by residents of group quarters (e.g., college dormitories) and the individual report filled by military and shipboard service members. The operation also included searches from the parolee and probationer program (described below) and the "Were You Counted?" program.

* Parolee/Probationer Program: the 1990 census added a special program for distributing "information record" forms to parole and probation officers, which in turn were to be distributed to each of their assignees. Each parolee or probationer was asked for their Census Day address as well as some basic demographic information. Due to lower-than-

expected response, the program was augmented in target counties expected to have large parolee or probationer populations; Census Bureau field staff directly contacted corrections departments to obtain administrative data, as available, on names and Census Day addresses.

– Coverage improvement techniques study: evaluation that attempted to quantify the coverage yields, financial costs, and errors associated with the numerous coverage improvement operations in the 1990 census.

– Coverage sampling research: experiment to estimate the feasibility of two options for reducing coverage error in the census: telephone reinterview of a sample of mailout-mailback households and reinterview (using more detailed household rostering questions) of a sample of nonresponse follow-up households. Samples were drawn in 52 urban mailout-mailback district offices (originally, the experiment was supposed to draw from 103 offices, but 51 were excluded due to their inclusion in other experimental studies or at the request of the Census Bureau's Field Division).

• *Evaluation of outreach and publicity:* set of studies to try to assess the effectiveness of the 1990 census outreach and advertising efforts. As in 1980, this included interviews with samples of households both prior to and immediately following Census Day; it also included structured focus-group discussions with representatives of community organizations and selected demographic subgroups.

SOURCE: Adapted from U.S. Census Bureau (1996:Chap. 11).

## A–6  2000 CENSUS

### A–6.a  Principal Pretests and Experiments Conducted Prior to the Census

### Pretests of Census Operations and Questionnaires

• *1992 (national sample):* Simplified Questionnaire Test (SQT) assessed the effects on mail return rates of form length and making the form more user-friendly; it used a prenotice letter, a reminder postcard (used in 1990), and a replacement questionnaire in all treatments. Five different short forms were tested: the booklet form used in 1990, which had been designed to be read by FOSDIC, as a control, and four user-friendly forms, including a booklet form, which contained all of the 1990 content; a micro form, which contained no housing items and asked only for name, age, sex, race, and ethnicity; the micro form with a request for Social Security number added; and a roster form, which asked only for name and date of birth. Both the micro and roster forms achieved higher return rates than the booklet form, especially in easy-to-enumerate areas; the user-friendly booklet form achieved a higher return rate than the control form, especially in hard-to-enumerate areas (Dillman et al., 1993a).

- *1993 (national sample):* Implementation Test (IT) assessed the effects on mail return rates of using a prenotice letter, a reminder postcard, a stamped return envelope, and a replacement questionnaire. All treatments used the micro short form. The stamped return envelope had no effect; the prenotice letter and reminder postcard increased return rates by 6 and 8 percentage points, respectively. From comparing the IT treatment that combined a prenotice letter and reminder postcard with the SQT for the micro form, the Census Bureau estimated that the use of a second questionnaire increased the return rate by 10–11 percentage points.

- *1993 (national sample):* Appeals and Long-Form Experiment (ALFE) assessed the effects on mail return rates of motivational appeals and of different formats of the long form. Three motivational appeals were tested with the booklet short form used in the SQT, emphasizing, respectively, the benefits of the census, the confidentiality of the data, and the mandatory nature of responding. The first two appeals had no effect; the third increased return rates by 10–11 percentage points. The long-form portion of the experiment compared the 20-page 1990 census long form, a 28-page user-friendly booklet format, and a 20-page user-friendly row-and-column format. The user-friendly booklet increased return rates compared with the 1990 long form, but only in easy-to-enumerate areas; the row-and-column format had no effect. The design of the experiment did not permit separating the effects of length from format for the long forms tested (Treat, 1993).

- *1993 (national sample):* Spanish Forms Availability Test showed that offering a Spanish-language form increased mail return rates by 2–6 percentage points; however, households that completed the Spanish form had higher item nonresponse rates than households that completed the English form, and most Hispanic households completed the English form.

- *March 1995 (Oakland, CA; Paterson, NJ; Bienville, De Soto, Jackson, Natchitoches, Red River, and Winn Parishes, LA):* "1995 Test Census" served as first major test of the Census Bureau's planned Integrated Coverage Measurement (ICM), use of an independent postenumeration survey to statistically adjust totals via dual-system estimation. New Haven, CT, was originally designated as a test site, but the Bureau decided in September 1994 to scrap that site due to budget considerations. Pursuant to the Census List Improvement Act of 1994, the 1995 Test Census was also the first opportunity for the Census

Bureau to test an address file updated using the U.S. Postal Service's Delivery Sequence File and a prototype Local Update of Census Addresses (LUCA) program (through which local officials were permitted to review housing unit addresses and maps) in the test jurisdictions. The Census Bureau also studied the extent to which mail returned as undeliverable actually corresponded to vacant or nonexistent units through a premailout postal carrier check similar to previous censuses and tests, through returned advance letters alerting households to the upcoming test, and through returned questionnaire packages. In terms of questionnaire design, the 1995 test used only one short-form design but tested several long-form alternatives ranging from 16 to 53 questions in order to assess differential response. The 1995 test also marked the first step toward what would later become known as service-based enumeration—the attempt to identify and locate homeless people through facilities where they receive services, rather than (or in addition to) nighttime deployment of enumerators, as done in previous censuses.

- *March–June 1996 (national sample of housing units from 1990 census mailback areas not already included in previous tests):* National Content Survey testing 13 alternative questionnaire forms, seven short-form designs and six long-form designs. Some treatments tested alternative structures for the race and Hispanic-origin questions (varying their order or allowing for multirace reporting); long-form variations also presented different structures for collecting employment and lay-off data (Raglin, 1998b).
- *June 1996 (Paterson, NJ):* field test of ICM procedures, referenced by Whitford (1996).
- *October 1996 (seven tracts in Chicago, IL; Pueblo of Acona Indian Reservation, NM; Fort Hall Reservation and trust lands, ID):* Dubbed the "1996 Community Census," the 1996 test served as a further demonstration of the Bureau's ICM procedures. At least in the Chicago test site, the independent postenumeration survey was conducted on a 100 percent basis (i.e., an interview was conducted at every household generated by an independent listing of addresses) rather than a sample basis, as planned in the census; this permitted fuller study of the overlap between the initial census and postenumeration survey coverages and facilitated comparison of consistency of reporting on questions like race and Hispanic origin. Also in Chicago, the Census Bureau constructed an administrative records database from numerous sources (including Internal Revenue Service data and the Social Security Administration's NUMIDENT file) for comparison with the test census results.

## Dress Rehearsals

In November 1997, a provision in P.L. 105–119 (Title II, § 209) directed that the Census Bureau "plan, test, and become prepared to implement a 2000 decennial census, without using statistical methods" (i.e., the Census Bureau's plans for ICM and sampling rather than following up all nonresponding households), and further required that both adjusted and unadjusted counts be made available in the 2000 census or "any dress rehearsal or other simulation made in preparation for the 2000 decennial census." As part of this compromise between the Census Bureau and Congress over fiscal year 1998 funding, the Bureau's planned dress rehearsal for 2008 in three sites was recast as a major operational test and comparison of alternative census designs.

- *2008 (Sacramento, CA):* full census operational trial using both ICM and sampling for nonresponse follow-up.
- *2008 (Columbia, SC, and 11 surrounding counties: Chester, Chesterfield, Darlington, Fairfield, Kershaw, Lancaster, Lee, Marlboro, Newberry, Richland, Union):* full census operational trial using 100 percent nonresponse follow-up; a postenumeration survey was conducted to assess coverage error but not to produce adjusted estimates.
- *2008 (Menominee County, WI, including the Menominee Indian Reservation):* full census operational trial using ICM and 100 percent nonresponse follow-up.

SOURCES: Adapted from Kim et al. (1998); National Research Council (1995, 1997, 2004a); Raglin (1998a,b); U.S. General Accounting Office (1994); Whitford (1996).

### A–6.b   Research, Experimentation, and Evaluation Program

## Experiments

- *Census 2000 Alternative Questionnaire Experiment (AQE2000):* this experiment used additional mailings and reinterview studies to manipulate three questionnaire design components:
    - Presentation of residence rules on the short form: determine whether providing a brief, reformatted version of the rules improves data quality.
    - Comparing the 1990 and 2000 census presentation of race and Hispanic-origin questions: the two censuses presented the questions in slightly different ways, in wording, format, and design.
    - Design of "skip to" and "go to" instructions in the census long form: determine whether respondents were able to navigate through the paper question correctly and efficiently.

Results are reported in Gerber et al. (2002); Martin (2002); Martin et al. (2003); Redline et al. (2002).

- *Administrative Records Census 2000 Experiment (AREX 2000):* initial planning for the 2000 census included experimentation with an administrative records census as a possible way to save costs; the idea had been raised but not endorsed by National Research Council (1995:Chap. 4) and National Research Council (1999:Chap. 5). The AREX 2000 experiment assembled national-level administrative records (unduplicated using SSNs) and assigned block-level geographic codes. Records for the five selected test sites were then extracted and tallied at the census block level. A separate branch of the experiment sought to reconcile administrative records with the Master Address File to generate block-level population and housing unit counts. The results of the experiment are reported in Bauder and Judson (2003); Berning (2003); Berning and Cook (2003); Heimovitz (2003); Judson and Bye (2003).

- *Social Security Number, Privacy Attitudes, and Notification Experiment (SPAN):* related to the administrative records research, the SPAN experiment probed for behavioral and attitudinal data on public response to queries for their SSNs on census questionnaires. The experiment also tested public response to variations in wording in notices about Census Bureau use of administrative records, as well as surveying public concerns about privacy and confidentiality raised by use of administrative records. The results of the experiment are reported in Brudvig (2003); Guarino et al. (2001); Trentham and Larwood (2003).

- *Response Mode and Incentive Experiment (RMIE):* this experiment studied the effectiveness of three electronic modes of data collection:

  - Operator telephone interview: also known as reverse computer-assisted telephone interview (reverse CATI); respondents were encouraged to call a toll-free telephone number, at which time a telephone interviewer administered the questionnaire.
  - Computer telephone interview: also known as the Automated Spoken Questionnaire; respondents were asked to call a toll-free telephone number, at which time the short-form questionnaire was administered by interactive voice response (an automated system).
  - Internet: Respondents were encouraged to answer the questionnaire using a web address provided in a cover letter.

The experiment also tested the impact on response of offering an incentive for completing the questionnaire (specifically, a telephone calling card valid for 30 minutes of free long-distance calling). The results are reported in Caspar (2003); Guarino (2001); Schneider et al. (2002).

- *Census 2000 Supplementary Survey (C2SS):* the C2SS extended pilot work on the American Community Survey (ACS). In addition to data collection in 36 ongoing ACS test sites, the C2SS collected data in 1,203 additional counties. The C2SS was conducted as an experiment, with the intent of determining whether it is feasible to collect long-form-census data at the same time, but in a separate process from, the decennial census data collection. The Census Bureau concluded that this simultaneous collection is feasible and that ACS work is feasible for a full national sample; the results are reported in Griffin and Obenski (2001).
- *Ethnographic studies*
    - Privacy Schemas and Data Collection: the goal of the experiment was to collect qualitative and attitudinal data on survey participation and response, including further probing of privacy concerns and elaborating reasons for choosing to participate in survey data collections. The results of the study are reported in Gerber (2003).
    - Complex Households and Relationships in the Decennial Census and Demographic Surveys: this ethnographic research project assembled six teams to study how well census methods, questions, and categories matched the diversity and experience of modern households. The six teams targeted particular ethnic or race groups: African Americans, Hispanics, Inupiaq Eskimos, Koreans, Navajos, and whites. The results are reported in Schwede (2003).
    - Generation X Speaks Out on Censuses, Surveys, and Civic Engagement: An Ethnographic Approach: this ethnographic study was intended to probe the civic engagement and attitude toward censuses (and surveys in general) among the Generation X population, those born between 1968 and 1979. In this age cohort, differences by other factors—socioeconomic background, ethnicity, immigrant status, and so forth—were also considered. Members of the subsequent Millennial generation (14–18 years of age) were also interviewed for comparison. The results are reported in Crowley (2003).

## Evaluations

The Census Bureau's original slate of evaluations for the 2000 census included 149 studies; in several waves during 2002, this list was "was refined and priorities reassessed due to resource constraints at the Census Bureau"

(U.S. Census Bureau, 2003), yielding a final set of 91 studies. However, some of the studies "cancelled" from the formal evaluation set were expedited and shifted to the research program surrounding the Executive Steering Committee for A.C.E. Policy, which was making recommendations on the adjustment of census figures for redistricting and other purposes.

- *Response Rates and Behavior Analysis (Series A)*

    - Telephone Questionnaire Assistance (TQA) Operational Analysis: assess calling patterns and respondent use of Telephone Questionnaire Assistance and performance of operational system.
    - TQA Customer Satisfaction Survey: present results of quality assurance survey administered to TQA respondents.
    - Internet Data Collection Operational Analysis: evaluate frequency of completion of short-form questionnaires via the Internet.
    - Internet Web Site and Questionnaire Customer Satisfaction Survey: present results of customer satisfaction surveys.
    - Be Counted Campaign: evaluate impact on coverage of Be Counted forms, blank questionnaires that were made publicly available for people who believed they had not been counted in the census.
    - Language Program (Use of Non-English Questionnaires and Guides): document requests for non-English census forms and language assistance guides.
    - Response Methods for Selected Language Groups: examine how certain non-English-speaking groups (Spanish, Chinese, Korean, Vietnamese, and Russian) were enumerated.
    - Awareness and Participation in the Language Assistance Programs Among Selected Language Groups: examine household awareness of Census Bureau's language assistance programs.
    - U.S. Postal Service Undeliverable Rates for Mailout Questionnaires: examine rate at which the Postal Service classified housing units as "undeliverable as addressed" and evaluate occupancy status of those units.
    - Detailed Reasons for Undeliverability of Census 2000 Mailout Questionnaires by the Postal Service: further analysis of reasons for questionnaires to some housing units being deemed "undeliverable as addressed," including follow-up work by local census offices.
    - Mailback Response Rates: examine mail response rates by geography and questionnaire check-in dates.
    - Mail Return Rates: examine mail return rates by geography,

and compare return rates based on housing unit demographic variables.

- Puerto Rico Focus Groups on Why Households Did Not Mail Back the Census 2000 Questionnaire: examine reasons for low response to update-leave enumeration as applied for the first time in Puerto Rico.
- *Cancelled:* Internet Questionnaire Assistance Operational Analysis.

• *Content and Data Quality (Series B)*

- Imputation Process for 100 Percent Household Population Items: examine national-level rates of assignments, allocations, and substitutions (using Census Bureau terminology).
- Item Nonresponse Rates for 100 Percent Household Population Items: compare item nonresponse rates by form type and response mode.
- Census Quality Survey to Evaluate Responses to the Census 2000 Question on Race: interpret results of a follow-up survey asking respondents to identify their race in two ways: choosing only one of several racial categories and choosing multiple. The objective is to use these results to help bridge 2000 census results (using multiple-race responses) with other data sources (where single-race responses may still prevail).
- Content Reinterview Survey: Accuracy of Data for Selected Population and Housing Characteristics as Measured by Reinterview: report results of a follow-up survey, asking long-form-census respondents to resupply long-form data items; the results from the census and survey are then compared for consistency.
- Master Trace Sample: archive and link the results for randomly selected census records across multiple operational databases, including address list information, data processing archives, enumerator information, and Accuracy and Coverage Evaluation data.
- Match Study of Current Population Survey to Census 2000: discuss results of person-level match between 2000 census returns and the Current Population Survey, emphasizing differences in estimates of poverty and labor force status.
- Puerto Rico Race and Ethnicity: examine responses to the race and Hispanic-origin questions from respondents living in Puerto Rico (the questions administered were identical to those asked in the rest of the nation).
- Puerto Rico Focus Groups on the Race and Ethnicity Questions:

results of focus groups conducted in Puerto Rico on perceptions of the race and Hispanic-origin questions.

- *Cancelled (6):* Documentation of Characteristics and Data Quality by Response Type; Match Study of A.C.E. to Census to Compare Consistency of Race and Hispanic-Origin Responses; Housing Measures Compared to the American Housing Survey; Housing Measures Compared to the Residential Finance Survey; ACS Evaluation of Follow-Up, Edits, and Imputations; Comparisons of Income, Poverty, and Unemployment Estimates Between Census 2000 and Three Census Demographic Surveys.

- *Data Products (Series C)*

  - Effects of Disclosure Limitation on Data Utility: evaluates refinements in the Census Bureau's use of "data swapping" to prevent inadvertent disclosure of confidential census information; in particular, the evaluation was intended to focus on the impact of a region's geographic structure and racial diversity on the effectiveness of data swapping.
  - *Cancelled (2):* Usability Evaluation of User Interface with American FactFinder; Data Products Strategy.

- *Partnership and Marketing (Series D)*

  - Partnership and Marketing Program: study conducted by the National Opinion Research Center of the effectiveness of the Census Bureau's marketing and advertising campaign.
  - Census in Schools/Teacher Customer Satisfaction Survey: study conducted by Macro International on the effect of offering Census-related educational materials to school teachers.
  - Survey of Partners/Partnership Evaluation: study conducted by Westat on the helpfulness of 2000 census materials provided to the Census Bureau's local and community organization partners.

- *Special Places and Group Quarters (Series E)*

  - Special Place/Group Quarters Facility Questionnaire (Mode Effect): evaluate effectiveness of the computer-assisted telephone interviewing (CATI) system or personal visit (PV) used to collect information on special places, based on personal visit reinterviews.
  - Decennial Frame of Group Quarters and Sources: evaluate content, coverage, and sources of the 2000 census roster of group quarters through comparison to other sources and business registers.
  - Group Quarters Enumeration: study aspects of group quarters enumeration, including counts of special places and group quar-

ters and distribution of group quarters per special place and residents per group quarters.

- Service-Based Enumeration: evaluate conduct of Service-Based Enumeration, which targeted populations in shelters, soup kitchens, regularly scheduled mobile food vans and targeted nonsheltered outdoor locations in March 2000.
- *Cancelled (3):* Special Place/Group Quarters Facility Questionnaire (Operational Analysis); Special Place LUCA; Inventory Development Process for Service-Based Enumeration.

• *Address List Development (Series F)*

- Address Listing Operation and Its Impact on the Master Address File (MAF): assess quality of address listing in areas to be enumerated via update-leave (but not geographic database updates made at the same time).
- LUCA 1998: evaluate operation that gave local and tribal governments opportunity to review address list entries in areas with predominantly city-style addresses, intended for mailout-mailback enumeration.
- Block Canvassing Operation: assess quality of 100 percent field canvass of mailout-mailback areas (but not geographic database updates made at the same time).
- LUCA 1999: evaluate operation that gave local and tribal governments opportunity to review block counts of housing units for areas with predominantly non-city-style addresses.
- Update-Leave: assess enumeration procedure in which mail delivery was deemed to be problematic and in which direct visits from enumerators were to be used.
- Urban Update-Leave: assess enumeration procedure used in selected areas in urban centers judged to be not amenable to mailout-mailback.
- Update-Enumerate: assess enumeration procedure through which interviewers visited housing units once, to verify address list entries and to collect questionnaire information, rather than simply dropping a questionnaire for later return by mail.
- List/Enumerate: assess enumeration procedure in which enumerators simultaneously listed housing units and collected questionnaire data.
- Quality of the Geocodes Associated with Census Addresses: study status of addresses still deemed to be "missing" from the census after A.C.E. work had finished.
- Block Splitting Operation for Tabulation Purposes: examine accuracy of block splitting operations, performed in some areas to

tabulate data when governmental unit boundaries do not conform to census collection block boundaries.

- *Cancelled (6):* Impact of the Delivery Sequence File Deliveries on the MAF; Evaluation of the MAF Using Earlier Evaluation Data; Criteria for the Initial Decennial MAF Delivery; Decennial MAF Update Rules; New Construction Adds; Overall MAF Building Process for Housing Units.

- *Field Recruiting and Management (Series G)*

  - Staffing Programs: eventually issued in two parts, the evaluation focused on hiring programs for nonresponse follow-up, examining both the adequacy of recruitment as well as the impact of higher pay rates on productivity.
  - *Cancelled:* Operation Control System.

- *Field Operations (Series H)*

  - Operational Analysis of Field Verification Operation for Non-ID Housing Units: report on the resolution of non-ID questionnaires—that is, those from the Be Counted or TQA operations (which were not keyed to a MAF ID) or questionnaires for which an address could not be verified.
  - Questionnaire Assistance Centers: report on effectiveness of walk-in assistance centers where respondents could be provided help in filling out the census questionnaire.
  - NRFU: principal evaluation of enumerator follow-up of nonresponding housing units, including determination of NRFU workloads, demographic profiles of the NRFU respondent population, and the distribution of partial interviews, refusals, and proxy responses.
  - NRFU Enumerator Training: assess the adequacy of training for the large temporary work corps hired to conduct nonresponse follow-up interviewing.
  - Operational Analysis of Enumeration of Puerto Rico: consider the effectiveness of enumeration in Puerto Rico, which was conducted using update-leave methodology in 2000.
  - Local Census Office Profile: operational summary of descriptive statistics on total housing units, average household size, mail return rate, and other information for each local census office.
  - Date of Reference for Respondents: examine discrepancies in respondents' reported age and their date of birth and derive the average date of reference actually used by census respondents (in comparison to the official comparison date of April 1, 2000).
  - *Cancelled (3):* Use of 1990 Data for Census 2000 Planning; Local Census Office Delivery of Census 2000 Mailout Questionnaires

Returned by U.S. Postal Service With Undeliverable as Addressed Designation; Operational Analysis of Non-Type of Enumeration Area Tool Kit Methods.

- *Coverage Improvement (Series I)*

    - Coverage Edit Follow-Up: assessment on program to resolve discrepancies between reported household size and actual number of people coded on the form, and to follow up on additional household members beyond the six household members whose date could be recorded on the standard questionnaire.
    - NRFU Whole Household Usual Home Elsewhere Probe: study resolution of cases in NRFU, List/Enumerate, and Update-Enumerate operations in which a respondent indicated that their current address was a seasonal or vacation home and their usual home was elsewhere. The evaluation was to consider how many such forms were filed and how well the reported "usual home" data compared with the MAF or other census enumerations.
    - NRFU Mover Probe: report on in-movers (people who moved to their current households after Census Day and detected in follow-up or direct enumeration methods), inquiring whether they completed a census form at their Census Day address.
    - Coverage Improvement Follow-Up: report on the phase of follow-up that verified and collected information from units flagged as vacant or delete earlier in follow-up; units added in the New Construction operation; and other late additions, blank mail returns, and lost mail returns.
    - Coverage Gain from Coverage Questions on Enumerator Completed Questionnaire: consider effectiveness of change in approach from 1990, obtaining information on missing or erroneously included persons through a set of questions rather than a recitation of residence definitions as in 1990.
    - *Cancelled:* Comparative Study of Coverage, Rostering Methods, and Household Composition in the Current Population Survey and the Census.

- *Ethnographic Studies (Series J)*

    - Ethnographic Social Network Tracing: use of social network analysis to study patterns of residential mobility and its impact on the census count, as well as how reliably people can be identified from their social networks.
    - Comparative Ethnographic Research on Mobile Populations: report on characteristics and challenges in enumerating select hard-to-count groups:urban gang members, Irish Travelers in Missis-

sippi and Georgia, Arizona "snowbirds," and American Indian populations in the San Francisco area.

- Colonias on the U.S./Mexico Border: ethnographic study of challenges (e.g., irregular housing stock, complex household structures, heightened concerns about confidentiality) in residential subdivisions along the U.S./Mexico border that are generally unincorporated, low income, and difficult to reach in survey research.

- *Data Capture (Series K)*
  - Data Capture Audit Resolution Process: document results of audit of data capture processing, including failure reasons, form type, and differential effects by data capture site.
  - Quality of the Data Capture System and the Impact of Questionnaire Capture and Processing on Data Quality: study impact of the automated data capture on data quality and compares data quality by the questionnaire item and form type, among other variables.
  - Impact of Data Capture Errors on Autocoding, Clerical Coding and Autocoding Referrals in Industry and Occupation Coding: study of use of automated data capture routines, as well as clerical review, in parsing reported industry and occupation information and coding into standardized categories.
  - *Cancelled (4):* Analysis of Data Capture System 2000 Keying Operations; Synthesis of Results from Data Capture Audit Studies; Analysis of the Interaction Between Aspects of Questionnaire Design, Printing, and Completeness with Data Capture; Performance of the Data Capture System 2000.

- *Processing Systems (Series L)*
  - Decennial Response File Stage 2 Linking and Setting of Expected Household Population: document the linkages drawn between census forms and returns during initial response file processing, as well as the accuracy of algorithms for calculating expected household size.
  - Operational Assessment of Primary Selection Algorithm Results: study of unduplication algorithm, used to resolve cases in which several questionnaires were received from the same address (MAF ID number).
  - Resolution of Multiple Census Returns Using Reinterview: further research regarding the accuracy of the Primary Selection Algorithm, using a follow-up reinterview on a sample of addresses affected by the algorithm.
  - Census Unedited File Creation: document results of process for

determining final housing unit inventory, done by merging information on the processed Decennial Response File (census returns) with the Decennial MAF.
- Beta Site: operational analysis of software evaluation facility within the Census Bureau that is responsible for integrating software systems of the census as well as to conduct security testing.
- *Cancelled:* Invalid Return Detection.

• *Quality Assurance Evaluations (Series M)*

- Evaluation of Quality Assurance Philosophy and Approach Used for the Address List Development and Enumeration Operations: document operational experiences with quality assurance approach.
- Effectiveness of Existing Variables in the Model Used to Detect Discrepancies During Reinterview, and the Identification of New Variables: document specific quality assurance measure used in cases in which enumerators' work suggested discrepancies.

• *A.C.E. Survey Operations (Series N)*

- Contamination of Census Data Collected in A.C.E. Blocks: study intended to determine success in keeping census and A.C.E. operations independent.
- Discrepant Results in A.C.E.: examine how quality assurance steps identified interviewers who entered discrepant data in the A.C.E. interview.
- Evaluation of Matching Error: assess error in the matching process used to identify missed or erroneously enumerated persons between the census and the A.C.E.
- Targeted Extended Search Block Cluster Analysis: study result of expending search for possible matches from adjoining blocks for all sample clusters (as in 1990) to a broader search (still targeted to clusters judged most likely to benefit from additional searching.
- Field Operations and Instruments for A.C.E.: assess quality of housing unit and person coverage in A.C.E. operations by examining the quality of the (independent) address listing, effect of follow-up interviewing, and noninterview rates.
- *Cancelled (16, most converted to separate A.C.E. evaluation program):* Analysis of Listing Future Construction and Multi-Units in Special Places; Analysis of Relisted Blocks; Analysis of Blocks With No Housing Unit Matching; Analysis of Blocks Sent Directly for Housing Unit Follow-Up; Analysis of Person Interview With Unresolved Housing Unit Status; Analysis on the Effects of Census Questionnaire Data Capture in A.C.E.; Analysis of

the Census Residence Questions Used in A.C.E.; Analysis of the Person Interview Process; Extended Roster Analysis; Matching Stages Analysis; Analysis of Unresolved Codes in Person Matching; Outlier Analysis in the A.C.E.; Impact of Targeted Extended Search; Effect of Late Census Data on Final Estimates; Group Quarters Analysis; Analysis of Mobile Homes.

- *Coverage Evaluations of the Census and of the A.C.E. Survey (Series O)*

    - Housing Unit Coverage Study: examine net coverage rate, gross omission rate, and erroneous enumeration rate of housing units, at various geographic levels as well as by A.C.E. poststrata.
    - Analysis of Conflicting Households: report on resolution of cases in A.C.E. housing unit matching when the census and the A.C.E. listed two entirely different families for the same unit.
    - Analysis of Proxy Data in the A.C.E.: study accuracy (match rates and erroneous enumeration rates) of data collected from proxy respondents: persons who are not members of the household, such as neighbors or landlords.
    - Housing Unit Duplication in Census 2000: study characteristics of duplicate housing units, attempting to identify census operations most likely to produce housing unit duplication.
    - Analysis of Deleted and Added Housing Units in the Census Measured by the A.C.E.: evaluate housing unit coverage on the early Decennial MAF.
    - Consistency of Census Estimates with Demographic Benchmarks: comparison of census results with demographic analysis benchmarks.
    - *Cancelled (20, most converted to separate A.C.E. evaluation program or combined with other evaluations):* Type of Enumeration Area Summary; Coverage of Housing Units in the Early Decennial MAF; P-Sample Nonmatches Analysis; Person Coverage in Puerto Rico; Housing Unit Coverage in Puerto Rico; Geocoding Error Analysis; E-Sample Erroneous Enumeration Analysis; Analysis of Nonmatches and Erroneous Enumerations Using Logistic Regression; Analysis of Various Household Types and Long-Form Variables; Measurement Error Reinterview Analysis; Impact of Housing Unit Coverage on Person Coverage Analysis; Person Duplication; Analysis of Households Removed Because Everyone in the Household Is Under 16 Years of Age; Synthesis of What We Know About Missed Census People; Implications of Net Census Undercount on Demographic Measures and Program Uses; Evaluation of Housing Units Coded as Erroneous Enumerations; Analysis of Insufficient Information for Matching and Follow-Up;

Evaluation of Lack of Balance and Geographic Errors Affecting Person Estimates; Mover Analysis; Analysis of Balancing in the Targeted Extended Search.

- *A.C.E. Survey Statistical Design and Estimation (Series P)*
  - *Cancelled (5, most converted to separate A.C.E. evaluation program):* Measurement of Bias and Uncertainty Associated with Application of the Missing Data Procedures; Synthetic Design Research/Correlation Bias; Variance of Dual System Estimates and Adjustment Factors; Overall Measures of A.C.E. Quality; Total Error Analysis.

- *Organization, Budget, and Management Information System (Series Q)*
  - Management Processes and Systems: study the Census Bureau's organizational structure and decision-making processes, as well as its interaction with the Census Monitoring Board, Congress, the General Accounting Office, and other outside interests.

- *Automation of Census Processes (Series R)*; reviews of technical systems conducted by Titan Corporation
  - Telephone Questionnaire Assistance: toll-free service provided by a commercial phone center to answer questions about the census and the questionnaire.
  - Coverage Edit Follow-Up: program to resolve count discrepancies and obtain missing data for large households.
  - Internet Questionnaire Assistance: system that allowed respondents to use the Census Bureau's Internet site to ask questions and receive answers about the census questionnaire or other census-related information.
  - Internet Data Collection: system that offered census short-form respondents the opportunity to respond via the Internet, using the 22-digit ID number found on their mailed census form.
  - Operations Control System 2000: system for control, tracking, and progress reporting for all field operations conducted for the census.
  - Laptop Computers for A.C.E.: systems used in A.C.E. follow-up interviewing.
  - A.C.E. Control System: tracking and control system behind the A.C.E. field operations.
  - Matching and Review Coding System for A.C.E.: system used to match A.C.E. returns to census records.
  - Pre-Appointment Management System/Automated Decennial Administrative Management System: system used for administrative management in the census, including tracking and processing of temporary enumerators, payroll, and background checks.

- American FactFinder: systems developed to provide access to census data via the Internet at http://factfinder.census.gov.
- Management Information System 2000: systems used to manage census operations, including tracking of dates and budgets and creation of progress reports of current status during census operations.
- Data Capture: systems developed for full electronic data capture and imaging of census questionnaires, using optical mark and optical character recognition.

SOURCE: Adapted from National Research Council (2004a:App. I).

## A–7 PRINCIPAL PRETESTS AND EXPERIMENTS CONDUCTED PRIOR TO THE 2010 CENSUS

### A–7.a Pretests of Census Operations and Questionnaires

- *2002 (Gloucester County, VA):* pilot testing of use of Census TIGER maps on a handheld computer (Pocket PC-class) by census interviewers. The tests involved only locating particular address features on the small-screen map, not using the computer map to navigate a route or the collection of GPS coordinates.
- *February–April 2003 (national sample):* the 2003 National Census Test was administered to a national sample of about 250,000 households, drawn from the set of households that was enumerated using mailout-mailback methodology in the 2000 census. Strictly mail-based, the 2003 test involved no field follow-up component. The test focused primarily on two issues:
  - Response mode and contact strategies: different experimental groups were offered the opportunity to reply by mail (traditional method), Internet, or interactive voice response (IVR, an automated telephone system). Groups also varied as to whether these response modes were offered as a choice or whether they were "pushed" (e.g., providing Internet directions but no actual paper questionnaire in the mailing). Finally, contact strategies (including targeted replacement questionnaires and reminder postcards) were also varied. This component of the test involved eight experimental groups, one with 20,000 households and the other seven with 10,000 households each.
  - Race and ethnicity (Hispanic-origin) question wording: seven treatment groups of 20,000 households each received different variations on the wording and arrangement of questions on race and Hispanic origin. Experimental settings included whether "some other race" was offered as a choice in the categories for

race, whether wording was slightly revised to ask respondents if they are "Hispanic" or if they are "of Hispanic origin," and whether instructions explicitly directed respondents to answer both questions.

The test was rounded out by a control group of 20,000 households; this group's questionnaire included the race and Hispanic-origin questions worded as they were in the 2000 census (unlike the 2000 census context, the 2003 test control group households were eligible for a replacement questionnaire in lieu of nonresponse follow-up).

The samples for all groups were stratified by response rate in the 2000 census, in which the classification was a grouping into "high" and "low" response groups based on a selected cut-off. Martin et al. (2003:11) comment that the low-response strata "included areas with high proportions of Blacks and Hispanics and renter-occupied housing units" and further comment that addresses in low-response areas were oversampled. Still, it is unclear whether the sample design generated enough coverage in Hispanic communities to facilitate conclusive comparisons—that is, whether it reached enough of a cross-section of the populace and a sufficiently heterogeneous mix of Hispanic nationalities and origins to gauge sensitivity to very slight and subtle changes in question wording.

With regard to the response mode and contact strategy portion of the test, results reported by Treat et al. (2003) suggest that multiple response mode options may change the distribution of responses by mode—shifting some would-be mail responses to Internet, for example. However, the addition of choices does not generally increase cooperation overall. The experience of the 2003 test suggests serious difficulties with the interactive voice response option; 17–22 percent of IVR attempts had to be transferred to a human agent when the system detected that the respondent was having difficulty progressing through the IVR questionnaire. Moreover, rates of item nonresponse were greater for IVR returns than for the (paper response) control group. Internet returns, by comparison, experienced higher item response rates than the control. As indicated in past research, reminder postcards and replacement questionnaires had a positive effect on response.

Martin et al. (2003) report that the race and Hispanic-origin question segment of the test showed mixed results. Predictably, elimination of "some other race" as a response category reduced "some other race" responses considerably, by 17.6 percent (i.e., Hispanic respondents apparently declined to write in a generic response like "Hispanic" or "other" if "some other race" was not a formal choice). The Bureau concluded that the 17.6 percent decline in generic race reporting "more than offset" the impact of a 6.4 percent increase in the estimated number of Hispanics declining to answer the race question altogether (Martin et al., 2003:15). Adding examples of ancestry groups (e.g., Salvadoran, Mexican, Japanese, Korean) boosted the reporting of detailed origins among Hispanics, Asians, and Native Hawaiian or Other Pacific Islanders. Treatment groups for which instructions were revised, instructing respondents to answer both the race and Hispanic-origin questions, produced the most puzzling results; levels of missing data on one or both questions increased, as did the percentage reporting themselves as Native Hawaiian or Other Pacific Islanders (relative to the control group).

• *February–July 2004 (7 neighborhoods of Queens County, NY [Astoria, Corona, Elmhurst, East Elmhurst, Jackson Heights, Long Island City, and part of Woodside]; Colquitt, Thomas, and Tift Counties, GA):* the 2004 test was intended to test a package of new procedures and technologies in both a high-density urban site and a rural site. Lake County, IL, was originally designated a test site but was dropped due to constraints in funding for fiscal year 2004. Intended to include approximately 200,000 housing units, the test centered around a few major topics:

  – Handheld devices: the test marked the Census Bureau's first attempt to use handheld computers, equipped with GPS receivers, for nonresponse follow-up interviewing. The handhelds developed for use in the 2004 test were pieced together from commercial, off-the-shelf components; although the test included some advanced workflows (e.g., transmitting questionnaire data directly from enumerators' devices to headquarters and pushing enumerator assignments directly to individual handhelds, without filtering through regional or local offices), some parts of case management and assignment were still done using paper reports.

  – Further testing of race and Hispanic-origin question wording: the 2004 test permitted continuation of work from the 2003 test.

  – Special place/group quarters definition: the 2004 test field exercises allowed Bureau staff to test the aptness of revised definitions of group quarters (nonhousehold) populations.

In addition to dropping the Illinois test site, other planned parts of the 2004 test were eliminated from the test plan, including an attempt to target a mailing of dual-language (English and Spanish) questionnaires to certain households and to target canvassing methods for updating the MAF.

The 2004 field test also provided the first chance to test a version of the Bureau's planned Coverage Follow-Up (CFU) operation, a combination of coverage improvement programs from previous censuses and computerized matching to detect potential census duplicates. Given the relatively small size of the test, the Bureau was able to perform follow-up on all eligible cases and to conduct a post hoc clerical review of cases to specify the likely source of duplication. Follow-up interviews were conducted both by telephone (experienced interviewers) and field staff (temporary, relatively novice interviewers).

- *July 2004 (France, Kuwait, and Mexico):* Overseas Enumeration Test meant to assess the feasibility of a count of all Americans living overseas, motivated by congressional interest. Interest in the issue had been heightened by one of Utah's legal challenges to the 2000 census (having been edged for the 435th seat in the U.S. House of Representatives by North Carolina) concerning the counting of missionaries of the Church of Jesus Christ of Latter-day Saints. To carry out the test, the Bureau adopted techniques similar to those used to elicit overseas resident counts in the 1960 and 1970 censuses: persons living overseas had to make contact with a U.S. embassy or consular office in order to obtain the census questionnaire. Some publicity about the test in the selected countries (France, Kuwait, and Mexico) was made in English-language newspapers and media. Costing approximately $8 million, the test yielded very low response: 5,390 questionnaires total, compared to 520,000 questionnaires printed for the test and rough estimates of on the order of 1.15 million American citizens in the test countries. Although the 2004 effort was originally intended to be repeated in 2006, no funding for the overseas test was included in the Bureau's appropriations for that year.

- *August–September 2005 (national sample):* a second, mail-only National Census Test in 2005 involved only variations in questionnaire design. The test reached a sample of about 420,000 households and was intended to simultaneously address several major objectives: (1) test revised questions on tenure, relationship, age, date of birth, race, and Hispanic origin; (2) test respondent friendliness of new designs of mail and Internet questionnaires; (3) compare revised versions of the Question 1 household count item, including residence rules instructions, and further test coverage probe questions selected from those tested in 2003; (4) test use of a replacement questionnaire; and (5) test use of a bilingual English-Spanish questionnaire. All of these objectives were intended to be covered by a set of 20 experimental treatment groups within a single-panel test; the control group included some items worded as they had been in the 2000 census, but other items used versions tested in 2003 or 2004. Although there was no field follow-up, a sample of respondents was reinterviewed by telephone to assess the consistency of household rostering. Logistically, the test encountered problems as it coincided with the impact of Hurricanes Katrina and Rita and the disruption or suspension of mail service in the Gulf Coast area; the test was also complicated by the decision to keep the Internet questionnaire English-only and in a single format that lacked the other experimental features, preventing Internet returns from being used in studying some treatments.

- *Early 2006 (national sample):* ad hoc Short Form Mail Experiment with a planned sample size of about 24,000 addresses from the U.S. Postal Service's Delivery Sequence File. Not originally part of the testing plan for the 2010 census, the Census Bureau sought authority for the test in a *Federal Register* notice in October 2005. As described in that notice (*Federal Register*, October 5, 2005, p. 58181), this special mailout test was developed based on three objectives:

  - "Evaluate the effects of the wording of the instruction about who to list as Person 1" on the questionnaire (the householder, with whom the reported relationships of other household members are defined);
  - "Evaluate the proportion of respondents who forget to enumerate themselves by asking them to provide their personal information at the end of the form" (with "personal information" described as name, phone number, and proxy status—that is, whether the respondent is completing a form for someone else); and
  - "Evaluate how a compressed schedule with a fixed due date impacts unit response patterns." (The notice specified only that the "compressed schedule" would change from questionnaires being "mailed 2 weeks before 'Census Day'" to "households receiv[ing] the questionnaires a few days before 'Census Day.'")

The Census Bureau divided the sample into four treatment groups (*Federal Register*, October 5, 2005, p. 58181):

  - Group 1. Housing units in this treatment group will receive questionnaires with the same wording for the Person 1 instruction that we used in the Census 2000 questionnaire. In the Final Question, respondents will be asked to provide their name, telephone number and proxy information. The mail out schedule will be the conventional schedule. The questionnaire will be mailed two weeks before "Census Day", and there will be no explicit deadline.
  - Group 2. Housing units in this treatment group will receive questionnaires with the revised wording for the Person 1 in-

struction. In the Final Question, respondents will be asked to provide their name, telephone number and proxy information. The mailout schedule will be the conventional schedule. The questionnaire will be mailed two weeks before "Census Day" and there will be no explicit deadline.

- Group 3. Housing units in this treatment group will receive questionnaires with the revised wording for the Person 1 instruction. In the Final Question, respondents will be asked to check over their answers before considering the survey complete. The mailout schedule will be the conventional schedule. The questionnaire will be mailed two weeks before "Census Day" and there will be no explicit deadline.

- Group 4. Housing units in this treatment group will receive questionnaires with the revised wording for the Person 1 instruction. In the Final Question, respondents will be asked to check over their answers before considering the survey complete. The mailout schedule will be compressed, so that the survey is received closer to "Census Day" and an explicit due date will be provided.

• *April 2006 (part of Travis County, TX, including the cities of Austin and Pflugerville; Cheyenne River Indian Reservation, SD):* a second operational test involving field work, the 2006 Census Test used mailout-mailback (with field nonresponse follow-up) in the Texas site and enumerator visits (no mail) in the South Dakota site. As in 2004, handheld computers were used for enumerator interviews and were assembled from off-the-shelf components; in 2006, the new aspect of handhelds being tested was the delivery of maps to enumerators via the devices. The Census Bureau's list of definitions of group quarters was expanded following the 2004 test, and the usefulness of this revised set was assessed in Travis County.

### A–7.b   Dress Rehearsal

• *April 2008 (San Joaquin County, CA; nine-county area surrounding Fayetteville, NC [Chatham, Cumberland, Harnett, Hoke, Lee, Montgomery, Moore, Richmond, and Scotland Counties, including Fort Bragg and Pope Air Force Base]):* as a full rehearsal of census activities, the 2008 dress rehearsal actually began operations in spring 2007 with address canvassing. This activity was the first operational trial of custom handheld devices designed by Harris Corporation and its subcontractors under the Field Data Collection Automation (FDCA) contract; problems encountered in use of the handhelds precipitated a crisis in funding and a "replan" of FDCA and core census operations in the first half of 2008.

SOURCES: Adapted from "2010 Census: How We Prepare for 2010" (http://2010.census.gov/2010census/about_2010_census/007623.html); for the 2006 Short Form Mail Experiment, *Federal Register*, October 5, 2005, pp. 58180–58182; Hill et al. (2006); Karl et al. (2005); Knight et al. (2005); National Research Council (2004b:Boxes 9.1 and 9.2); Pennington (2005); Tancreto (2006).

# – B –

# 2010 Census Program of Evaluations and Experiments

Since the issuance of the panel's letter report in February 2009, some additional detail about the structure of the 2010 Census Program of Experiments and Evaluations (CPEX) has become available. Some revisions to specific CPEX components were indicated in a formal response by the Census Bureau to the panel's letter report (U.S. Census Bureau, 2009a). Other information and detail on planned CPEX components have also been made available because of the Census Bureau's filing for general approval of CPEX data collection with the U.S. Office of Management and Budget (OMB). Under law, OMB is responsible for reviewing and approving any information collection activity that will be administered to 10 or more respondents.[1] The Census Bureau requested generic clearance of parts of the CPEX program involving original data collection,[2] and the request was approved on May 12, 2009; subsequent specific changes and project plans have been submitted in addition to this generic clearance. The package (Information

---

[1] As part of the clearance process, agencies must have also made two postings in the *Federal Register* to solicit public comment. Packages submitted by agencies must be accompanied by two supporting statements: Part A, giving a detailed justification of the collection and indicating how the data will be used, and—for statistical collections—Part B on methodology, sampling strategy, and preliminary testing.

[2] "Generic clearance" seeks general authority to conduct data collection up to some set number of respondent-burden hours; though the content and questions of the data collection must still be specified, agencies have more latitude to make changes and revisions to specific forms and questions under a generic clearance.

---

**Box B-1**   Experiments in the 2010 CPEX

- *Alternative Questionnaire Experiment*
- *Deadline Messaging Experiment*
- *Nonresponse Follow-Up Contact Strategy Experiment*
- *Privacy Notification Experiment*
- *"Heavy-Up" Communications Experiment*

SOURCES: Presentations to the panel; "2010 CPEX Information Sheet: 2010 ICP Paid
Media Heavy-up Experiment," shared with panel in May 2008.

---

Collection Review [ICR] 200902-0607-007) is accessible through OMB's
RegInfo web site (http://www.reginfo.gov).

## B–1   EXPERIMENTS

As presented to the panel in early 2009, the Census Bureau's 2010 CPEX
program included four formal experiments. Since then, a fifth experiment
has been added to the ranks. Box B-1 lists the experiments for ease of reference; we provide additional description (and extend commentary from our
letter report, as appropriate) on the experiments in the remainder of this
section.

### B–1.a   Alternative Questionnaire Experiment

An Alternative Questionnaire Experiment (AQE) in which a sample of
census respondents receives questionnaires that vary in content, layout, and
question ordering and wording has been a staple of census experimentation
since the 1950 census. In that census, 10 district offices in Ohio and Michigan were used as "experimental areas" in which—among other things—four
census forms were oriented toward households as the unit of analysis and
self-response by individuals, as opposed to the person-based ledgers then
used by enumerators in conducting their interviewers (U.S. Census Bureau,
1955:5). The 2000 census AQE focused heavily on the effect of visual cues
and narrative instructions to guide respondents through the census long-
form questionnaire.  It also included an experimental group that varied
the instructions and formatting of the basic residence (household count)
question on the census form; the National Research Council (2006:202–
203) observed that this single treatment constituted "a bundle of at least 10
changes," some major and others extremely subtle, that rendered it impossible to ferret out which features were more or less effective than others.

The 2010 AQE is planned to include 19 panels, most of which (15) involve variations to the questions on race and Hispanic origin. The experi-

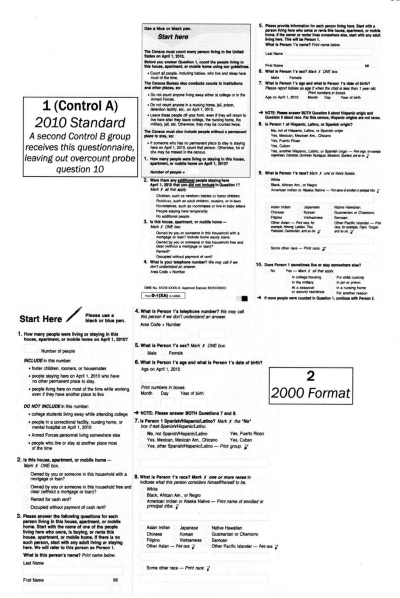

**Figure B-1** 2010 Alternative Questionnaire Experiment, control questionnaires

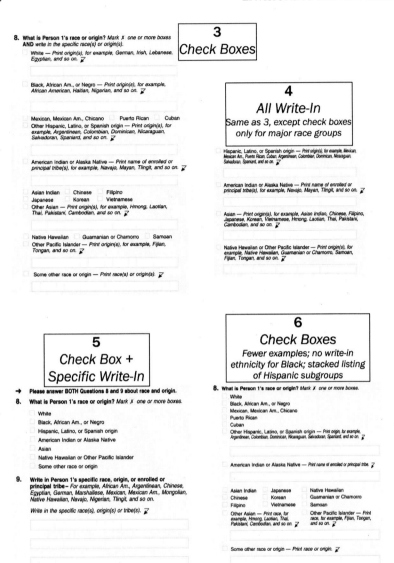

**Figure B-2**  2010 Alternative Questionnaire Experiment, structures of combined race and Hispanic origin question

→ NOTE: Please answer BOTH Question 8 about Hispanic origin and
Question 9 about race. For this census, Hispanic origins are not races.

**8. Is Person 1 of Hispanic, Latino, or Spanish origin?**

☐ **No,** not of Hispanic, Latino, or Spanish origin
☐ Yes, Mexican, Mexican Am., Chicano
☐ Yes, Puerto Rican
☐ Yes, Cuban
☐ Yes, another Hispanic, Latino, or Spanish origin — *Print origin, for example, Argentinean, Colombian, Dominican, Nicaraguan, Salvadoran, Spaniard, and so on.* ↗

**9. What is Person 1's race?** *Mark X one or more boxes.*

☐ White — *For example, German, Irish, Lebanese, Egyptian, and so on.*
☐ Black, African Am., or Negro — *For example, African American, Haitian, Nigerian, and so on.*
☐ American Indian or Alaska Native — *Print name of enrolled or principal tribe, for example, Navajo, Mayan, Tlingit, and so on.* ↗

☐ Asian Indian    ☐ Japanese    ☐ Native Hawaiian
☐ Chinese         ☐ Korean      ☐ Guamanian or Chamorro
☐ Filipino        ☐ Vietnamese  ☐ Samoan
☐ Other Asian — *Print race, for example, Cambodian, Pakastini, Mongolian, and so on.* ↗   ☐ Other Pacific Islander — *Print race, for example, Tongan, Fijian, Marshallese, and so on.* ↗

☐ Some other race — *Print race.* ↗

───────────────

**9. What is Person 1's race?** *Mark X one or more boxes.*

☐ White
☐ Black, African Am., or Negro
☐ American Indian or Alaska Native — *Print name of enrolled or principal tribe.* ↗

        <u>Asian</u>                    Native Hawaiian and
                                       Other Pacific Islander

☐ Asian Indian    ☐ Japanese    ☐ Native Hawaiian
☐ Chinese         ☐ Korean      ☐ Guamanian or Chamorro
☐ Filipino        ☐ Vietnamese  ☐ Samoan
☐ Other Asian — *Print for example, Cambodian, Hmong, Laotian, Pakistani, Thai, and so on.* ↗   ☐ Other Pacific Islander — *Print for example, Fijian, Tongan, and so on.* ↗

☐ Some other race — *Print race.* ↗

───────────────

| **7** |
| *Revised Examples* |
| Uses "Hispanic origins are not races" NOTE from control; revises examples for Other Asian and Other Pacific Islander |

| **17** |
| *Drop "Race"* |
| Same as 7, but omits mention of word "race" in NOTE and question wording |

→ NOTE: Please answer BOTH Question 8 and Question 9.

**9. Is Person 1...** *Mark X one or more boxes.*

| **15** |
| *Group Headings, Fewer Examples* |
| Otherwise same as 7, including NOTE |

| **16** |
| *Asian Examples* |
| Otherwise same as 15 |
| Other Asian — *Print race, for example, Hmong, Laotian, Thai, Pakistani, Cambodian, and so on.* ↗ |

| **18** |
| *Drop "Race"* |
| Same as 15, but omits mention of word "race" (as in 17) and uses Other Asian examples from 16 |

**Figure B-3**  2010 Alternative Questionnaire Experiment, variations on race question

→ **NOTE: Please answer BOTH Question 8 about Hispanic origin and Question 9 about race. For this census, Hispanic origins are not races.**

8. **Is Person 1 of Hispanic, Latino, or Spanish origin?**

☐ **No,** not of Hispanic, Latino, or Spanish origin
☐ Yes, Mexican, Mexican Am., Chicano
☐ Yes, Puerto Rican
☐ Yes, Cuban
☐ Yes, another Hispanic, Latino, or Spanish origin — *Print origin, for example, Dominican, Salvadoran, Colombian, Spaniard, and so on.* ↗

9. **What is Person 1's race?** *Mark X one or more boxes.*

☐ White
☐ Black or African Am.
☐ American Indian or Alaska Native — *Print name of enrolled or principal tribe.*

☐ Asian Indian    ☐ Japanese       ☐ Native Hawaiian
☐ Chinese         ☐ Korean         ☐ Guamanian or Chamorro
☐ Filipino        ☐ Vietnamese     ☐ Samoan
☐ Other Asian — *Print race, for example, Hmong, Laotian, Thai, Pakistani, Cambodian, and so on.* ↗    ☐ Other Pacific Islander — *Print race, for example, Fijian, Tongan, and so on.* ↗

☐ Some other race — *Print race.* ↗

---

**8**

**Revised Examples**

*Uses "Hispanic origins are not races" NOTE from control; revises examples for Other Hispanic origin*

---

8. **Is Person 1 of Hispanic, Latino, or Spanish origin?** *Mark X one or more boxes.*

☐ **No,** not of Hispanic, Latino, or Spanish origin
☐ Yes, Mexican, Mexican Am., Chicano
☐ Yes, Puerto Rican
☐ Yes, Cuban
☐ Yes, another Hispanic, Latino, or Spanish origin — *Print one or more origins, for example, Argentinean, Colombian, Dominican, Nicaraguan, Salvadoran, Spaniard, and so on.* ↗

---

**9**

**Multiple Hispanic Choices**

*Same as control, but adds "Mark one or more" instruction*

---

**10**

**Multiple Hispanic Choices/ Revised Examples**

*Uses Other Asian, Pacific Islander, and Hispanic examples from 7 from 8 and "Mark one or more" instruction from 9*

---

**11**

**Revised Examples**

*Uses Other Asian, Pacific Islander, and Hispanic examples from 7 from 8, but not "Mark one or more" instruction from 9*

---

**12**

**Multiple Hispanic Choices/ Revised Examples**

*Uses Other Asian and Pacific Islander examples from 7; adds "Mark one or more" instruction from 9*

---

**13**

**Multiple Hispanic Choices/ Revised Examples**

*Uses Other Hispanic examples from 8 and "Mark one or more" instruction from 9*

---

**Figure B-4**  2010 Alternative Questionnaire Experiment, variations on Hispanic question and hybrid approaches

**10.** Does Person 1 sometimes live or stay somewhere else?

☐ No →*SKIP to Person 2, if more people live here.*

☐ Yes — *Mark* ✗ *all that apply.*

| | |
|---|---|
| ☐ In college housing | ☐ At a seasonal or second residence |
| ☐ In the military | ☐ In jail or prison |
| ☐ For a job or business | ☐ In a nursing home |
| ☐ For child custody | ☐ For another reason |

**11.** IF you marked yes to Question 10, please provide the full address of the **other place** where Person 1 sometimes lives or stays:

House Number

Street Name

Apartment Number

Rural Route Address

City

State     ZIP Code

County

➡ NOTE: If there is no street address or if this is a facility, please print a description in the boxes below.

**12.** Where does Person 1 live or stay most of the time?

☐ The address printed on the back of this questionnaire
☐ The address or location you listed in Question 11
☐ Both places equally
☐ Some other place

**13.** On April 1, 2010, where was Person 1 staying?

☐ The address printed on the back of this questionnaire
☐ The address or location you listed in Question 11
☐ Some other place

---

**14**

*Other Residence*

*Asks for alternative address information based on "Yes" response to overcount probe question 10; such addresses also collected for Persons 2–6, requiring printing as a booklet*

---

**Figure B-5**  2010 Alternative Questionnaire Experiment, other residence panel mental panels are as follows and are numbered in the accompanying figures using the Census Bureau's scheme:

- *Controls:* The AQE shares a planned control group of 30,000 housing units with several of the other experiments described below; this group receives the standard 2010 census questionnaire with the only difference being that the phone number listed on the form for respondents to call if they have questions is a special CPEX line. However, the AQE includes a second control group that omits the overcount coverage probe question (i.e., "Does Person 1 sometimes live or stay somewhere else?") on the 2010 census form because the space requirements of the revised race and Hispanic-origin questions in the other experimental panels precluded the overcount question from fitting on the form.[3]

- *Cumulative Changes from 2010:* As shown in Figure B-1, one of the treatment groups uses the format and wording of the 2000 census questionnaire (save that it uses the 2010-standard blue color scheme rather than 2000's yellow color). The 2000 census form did not include either the overcount or undercount ("Were there any *additional* people staying here [on] April 1, 2010 that you *did not include* in Question 1?") coverage probe questions that have been added for 2010, and so those questions are not included on the 2000-style form. This treatment group is meant for comparison with the control group in order to study the cumulative effect of design and format changes from 2000 to 2010, although it will not be able to shed light on which specific features were more or less effective than others.

- *Combined Race and Hispanic-Origin Question:* Four panels test possible structures for a "combined" race and Hispanic-origin question that lists Hispanic origin (and related subgroups) in line with traditional race categories, as shown in Figure B-2. Two of the treatments (numbers 4 and 5) limit check-box choices to major race categories (permitting write-in of specific origins, nationalities, or tribal affiliations), and the others add specific check boxes for selected subgroups.

- *Variations on the Race Question:* As illustrated in Figure B-3, five experimental panels reflect suggestions from the Census Bureau's Race and Ethnic Advisory Committees, varying the mix of listed examples for Other Asian (e.g., omitting "Thai" and adding "Mongolian") and

---

[3]A consequence of the overcount question not being included in most of the panels of the AQE, given that the overcount question is one of the ways households are flagged for inclusion in the Coverage Follow-Up (CFU) operation, is that the AQE households are less likely to enter into the CFU workload. The Census Bureau estimates that this creates "the possibility of about 1,600 individuals being double-counted in the Census" because their households are not entered into CFU.

Other Pacific Islander (adding "Marshallese") groups. Two of the treatment groups (17 and 18) omit the word "race" from the wording of Question 9 and from the prefatory note, using the word only in the "Some other race" category.[4]

- *Variations on the Hispanic-Origin Question:* Also based on input from the Race and Ethnic Advisory Committee, one treatment group varies the listed examples of Hispanic subgroups (group 8) while another explicitly permits respondents to check more than one Hispanic category, as is permitted for the race question (group 9; see Figure B-4).

- *Joint Variations of Race and Hispanic Origin:* Also listed in Figure B-4, four treatment groups test combinations of the revised lists of race and Hispanic-origin examples with the "mark one or more" instruction on Hispanic origin.

- *Other Residence Information:* As shown in Figure B-5, a final experimental group in the AQE expands on the overcount coverage probe question. If the respondent indicates that a person sometimes lives or stays somewhere else, he or she is prompted to provide address information for that other location. Two follow-up questions then attempt to determine which address is the "usual" residence ("live or stay most of the time") and which was the "current" residence on Census Day.

Originally planned for a target sample size of 560,000 households—30,000 per treatment group except for the 2000-format panel 2, with 20,000 households—the exact numbers in each group may vary because of a change in approach to drawing the experimental sample. In previous censuses, experimental groups were drawn once, from strata defined at the national level. However, the Census Bureau notes that the 2010 AQE will be different (ICR 200902-0607-007, individual request for clearance document for the AQE):

> For the 2010 processing system development, schedule and time limitations necessitate sampling each Local Census Office (LCO) before moving onto the next, rather than selecting our sample once at the national level. This means that, since we will not have universe totals by stratum during the LCO sample selection, we have to fix our sampling intervals and let sample sizes vary.

As a result of this process, the Census Bureau estimates that the actual sample size of each 30,000-unit panel will be between 23,000 and 40,000 (and

---

[4]In using the phrase "some other race"—even in an experimental treatment—the Census Bureau is complying with a congressional mandate. A passage in census appropriations language directs "that none of the funds provided in this or any other Act for any fiscal year may be used for the collection of census data on race identification that does not include 'some other race' as a category" (P.L. 110-161; 121 Stat. 1887). This provision was first inserted into the Census Bureau's appropriations for fiscal year 2005, in an omnibus spending bill enacted in December 2004.

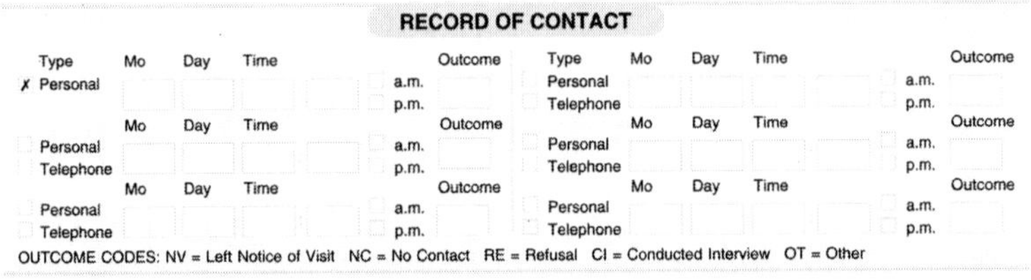
RECORD OF CONTACT

| Type | Mo | Day | Time | | Outcome | Type | Mo | Day | Time | | Outcome |
|---|---|---|---|---|---|---|---|---|---|---|---|
| ✗ Personal | | | | a.m. p.m. | | Personal Telephone | | | | a.m. p.m. | |
| Personal Telephone | Mo | Day | Time | a.m. p.m. | Outcome | Personal Telephone | Mo | Day | Time | a.m. p.m. | Outcome |
| Personal Telephone | Mo | Day | Time | a.m. p.m. | Outcome | Personal Telephone | Mo | Day | Time | a.m. p.m. | Outcome |

OUTCOME CODES: NV = Left Notice of Visit NC = No Contact RE = Refusal CI = Conducted Interview OT = Other

**Figure B-6** 2010 nonresponse follow-up enumerator questionnaire, record of contact box

15,000–26,000 for the 2000-format panel). The sample for the race and Hispanic-origin groups will be constructed hierarchically to try to ensure that the smallest demographic groups of interest are in the sample. In the ideal, fixed sample-size case, this would involve first drawing 9,000 households per panel from tracts with 15 percent or more Asian or Pacific Islander people, then 9,000 from tracts with 25 percent or more black people, then from tracts with 40 percent or more Hispanic people, and finally 3,000 from all other tracts. For the other residence information panel, the Bureau indicated to OMB that it will attempt to target tracts with high densities of active-duty military personnel, seniors (possible nursing home residents), college-age students, and "areas more likely to have child custody coverage issues," although how this will be done is not specified.

The AQE is also a factor in one of the formal evaluations in the CPEX program—a reinterview study—as described below in Section B–2.a.

### B–1.b Nonresponse Follow-Up Contact Strategy Experiment

The 2010 census will follow the example of recent censuses by calling for temporary enumerators to attempt up to six contacts with households in the nonresponse follow-up (NRFU) workload. Three of these visits are supposed to be in-person contact attempts, and the other three can be done by telephone if a phone number is available. If no contact is made by the sixth try, then proxy information (i.e., from a landlord or neighbor) can be collected as a last resort. The contact log section of the paper-based enumerator questionnaire planned for use in 2010 is illustrated by Figure B-6.

The Nonresponse Follow-Up Contact Strategy Experiment of the 2010 CPEX proposes to test the effect of reducing the maximum number of contacts to either 4 or 5. As described to the panel in November 2008, the experiment was to take place in three local census offices and a total of about 52,000 housing units; all enumerators would use the same form with space for six contact attempts, but enumerators in sampled crew leader districts would receive different instructions on how many callbacks to make.

The Bureau argued that logistical challenges in varying field procedures (and training for fieldwork) precluded a larger sample of offices or housing units. We criticized the experiment in the letter report, principally on the grounds that it would be extremely difficult to generalize data from only three offices.

Subsequently, the Bureau designed two alternative enumerator questionnaires—each of which uses the same amount of space for the contact log as in Figure B-6 but simply omits one or two entries (and adjusts the layout so that the remaining entries fill the space). As described in the Bureau's reply to our letter report (U.S. Census Bureau, 2009a) and the Bureau's filing with OMB, the design now calls for 1.2 million of these experimental enumerator forms (600,000 for each of the 4-contact- and 5-contact-maximum groups) to be randomly inserted with the standard enumerator forms, and hence for the experiment to take place in all 494 local census offices. The 1.2 million sample size is said to be large enough so that, on average, every enumerator will have approximately one experimental questionnaire in each of their assignment areas. Although the sample size is now massive, the redesigned experiment has the drawback that the enumerators will be acutely aware of the maximum number of attempts they can make; arguably, then, the test is less about the effect of simply cutting off NRFU after (say) four attempts than it is about whether enumerators will expend extra effort to try to resolve cases in four tries (in the same way that they may try particularly hard to make contact on the sixth time in normal cases).

### B–1.c Deadline Messaging/Compressed Schedule Experiment

The Deadline Messaging/Compressed Schedule Experiment is intended to see whether mail response (and speed of response) improves when different messages urging a rapid response are included in four mailing pieces:

1. the advance letter sent prior to the main questionnaire mailout,

2. the envelope containing the census form,

3. the cover letter accompanying the census form, and

4. the reminder postcard sent after the main mailout but before beginning nonresponse follow-up.

As originally described to the panel in November 2008, the experiment included three types of deadline messages; subsequently, a fourth has been added. The four message types are:

1. "Mild," which simply suggests a date by which the form should be returned;

2. "Progressive Urgency," which casts the date as a deadline for response and reminds the respondent that response is required by law;

3. "Avoid NRFU Visit" (or, as the Bureau refers to it, "NRFU Motivation"), which casts the date as a deadline and urges the respondent to avoid the trouble of having a nonresponse follow-up interviewer visit their home; and

4. "Cost Savings," which notes that money is saved by simply mailing the census form rather than having an interviewer come to visit.

It is useful to note that two of these message types repeat language used in 1990 census materials but not in 2000. The "Avoid NRFU Visit" message recalls a statement made directly in the instructions on the 1990 census form: "Avoid the inconvenience of having a census taker visit your home." Likewise, the "Your Guide to the 1990 U.S. Census Form" brochure distributed with the 1990 census included a statement very similar to the "Cost Savings" message: "If you do not mail back your census form, a census taker will be sent out to assist you. But it saves time and your taxpayer dollars if you fill out the form yourself and mail it back."[5] The manner in which these deadline messages are rendered in the mailed items is shown in Table B-1.

In the experiment, each of the deadline message strategies is used in combination with one of two schedules. Following the normal 2010 census schedule, advance letters are supposed to arrive between March 8 and March 10, the census questionnaire between March 15 and March 17, and the reminder postcards between March 22 and March 24. The alternative, "compressed" schedule shifts these dates by one week, closer to the April 1 Census Day: that is, advance letters arriving March 15–17, questionnaires March 22–24, and postcards March 29–31. Under the compressed schedule, the target or deadline date of April 5 referenced in Table B-1 remains the same.

In our letter report, we offered little comment on the Deadline Messaging/Compressed Schedule Experiment, noting only that—like the other experiments—we were concerned about the lack of an analysis of the statistical power of the proposed test to discriminate between the alternatives. At the time, as presented to us in November 2008, each study panel was to include 10,000 households (drawn from sampling strata of expected high, medium, and low mail response based on the 2000 census). In replying to our letter (U.S. Census Bureau, 2009a), the Census Bureau cited two unpublished Census Bureau internal memoranda (not shared with the panel) as justifying the sample size selection; the reply also indicated that the number of housing units per panel in the deadline experiment would be doubled to 20,000, without any indication of how this level was determined.

---

[5] However, the 1990 census materials also explicitly and repeatedly directed respondents to "return the completed form by April 1, 1990"—that is, to report information on one's household as of Census Day before that day had actually arrived—which the National Research Council (2006:Sec. 6–F) notes is a basic "violation by design" of the underlying census residence concept.

**Table B-1** 2010 Deadline Messaging/Compressed Schedule Experiment, deadline message treatments by form type

| Panel | Advance Letter | Initial Mailing Envelope | Cover Letter | Reminder Postcard |
|---|---|---|---|---|
| Control (2010 Standard) | When you receive your form, please fill it out and mail it in promptly. | YOUR RESPONSE IS REQUIRED BY LAW | Please complete and mail back the enclosed census form today. | If you have not responded, please provide your information as soon as possible. |
| 1 (Mild) | When you receive your form, please fill it out and mail it in by April 5. | YOUR RESPONSE IS REQUIRED BY LAW ¶ Mail by April 5 | Please complete and mail back the enclosed census form by April 5. | If you have not responded, please provide your information by April 5. |
| 2 (Progressive Urgency) | When you receive your form, please fill it out and mail it in by April 5. | YOUR RESPONSE IS REQUIRED BY LAW ¶ Deadline is April 5 | The deadline to complete and mail back the enclosed census form is April 5. | If you have not responded, the deadline to provide your information is April 5. Your response is required by law. |
| 3 (Avoid NRFU Visit) | When you receive your form, please fill it out and mail it in by April 5. | YOUR RESPONSE IS REQUIRED BY LAW ¶ Mail by April 5 | Please complete and mail back the enclosed census form by April 5 so that you can avoid a personal visit from an interviewer. | If you have not responded, please provide your information by April 5 so that you can avoid a personal visit from an interviewer. |
| 4 (Cost Savings) | When you receive your form, please fill it out and mail it in by April 5. | YOUR RESPONSE IS REQUIRED BY LAW ¶ Mail by April 5 | Please complete and mail back the enclosed census form by April 5. Mailing your census form on time saves money that would otherwise be used to follow up with you. | If you have not responded, please provide your information by April 5. Mailing your census form on time saves money that would otherwise be used to follow up with you. |

### B–1.d   Confidentiality/Privacy Notification Experiment

The cover letter accompanying a decennial census questionnaire typically is a letter or other printed message (often over the signature of the director of the Census Bureau) that assures respondents that the information they provide is kept confidential and is not disclosed to other government agencies. The planned cover letter for the 2010 census includes a short letter on one side and a short statement on confidentiality on the reverse side (the letter suggests that the document be turned over to read that information). The Confidentiality/Privacy Notification Experiment of the CPEX program involves two experimental treatments that make small changes to both the front and back of the cover letter, as shown in Figure B-7; these small changes are also made in the text of the letter that accompanies the second (replacement) questionnaire that the 2010 census will mail prior to the start of nonresponse follow-up. A control group—shared with the AQE and deadline messaging experiment—receives the standard 2010 cover letter in both the initial and second questionnaire mailings. A relatively small exercise as such census experiments go, the Confidentiality/Privacy Notification Experiment is planned to include 20,000 housing units in each of the two experimental panels.

### B–1.e   Heavy-Up Publicity Experiment

In May 2009, the panel was informed that a fifth experiment—a "heavy-up" test of increased media buys for census publicity materials in selected markets—had been added to the CPEX program. The experiment is apparently intended to study the effectiveness of saturated advertising (base messages via local media buys or culturally/ethnically targeted media buys) in selected marketing areas. Because the only information the panel has heard or seen concerning this test is a short "fact sheet," and the heavy-up experiment is not referenced in the Bureau's CPEX clearance request to OMB, we cannot provide additional commentary on this experiment akin to what we provided for the other four CPEX experiments in our letter report.

### B–2   EVALUATIONS

The Census Bureau's supporting statement for its OMB submission for CPEX (ICR 200902-0607-007) describes it as including "over 20 evaluations." Box B-2 lists the evaluation topic areas and the names of individual studies as they were presented to the panel in early 2009. The OMB submission provides detail on four of the proposed evaluations that involve original data collection, as discussed below.

Dear Resident:

This is your official 2010 Census form. We need your help to count everyone in the United States by providing basic information about all the people living in this house or apartment. **Please complete and mail back the enclosed census form today.**

Your answers are important. Census results are used to decide the number of representatives each state has in the U.S. Congress. The amount of government money your neighborhood receives also depends on these answers. That money is used for services for children and the elderly, roads, and many other local needs.

Your answers are confidential. This means the Census Bureau cannot give out information that identifies you or your household. Your answers will only be used for statistical purposes, and no other purpose. The back of this letter contains more information about protecting your data.

Sincerely,

*(reverse side)*

**Your Answers Are Confidential**

Federal law protects your privacy and keeps your answers confidential (Title 13, United States Code, Sections 9 and 214). The answers you give on the census form cannot be obtained by law enforcement or tax collection agencies. Your answers cannot be used in court. They cannot be obtained with a Freedom of Information Act (FOIA) request.

As allowed by law, census data becomes public after 72 years (Title 44, United States Code, Section 2108). This information can be used for family history and other types of historical research.

Please visit our Web site at <www.census.gov/2010census> and click on "Protecting Your Answers" to learn more about our privacy policy and data protection.

### Control

*Same "Your answers are confidential" paragraph used on cover letter accompanying replacement questionnaire*

### Privacy 2

*Revise "Your answers will only be used" sentence*

Your answers are confidential. This means the Census Bureau cannot give out information that identifies you or your household. Your answers will only be used to produce statistics. The back of back of this letter contains more information about protecting your data.

### Privacy 1 and 2

*Insert paragraph on back*

To improve census results, other government agencies may give us additional information about your household. The additional information we receive is legally protected under Title 13, just like your census answers.

**Figure B-7**  2010 Confidentiality/Privacy Notification Experiment, control and experimental treatments

---

**Box B-2**  Evaluations in the 2010 CPEX

**Coverage Improvement**

- Address frame accuracy—in preliminary design, said to involve (a) comparison of census coverage measurement (postenumeration survey) with census returns to study address errors and (b) additional field data collection in discrepant cases to establish "ground truth"
- Address canvassing targeting—analysis of actual address canvassing results in a set of blocks identified before the operation as "growth" areas
- Data-based extraction processes for address frame—use of data mining techniques to derive decision trees to predict address validity from address source(s), validated by comparison with address canvassing results
- Small multiunit structures—in preliminary design, said to include assessment of revised training materials and use of a special field activity to study possible identifiers for such structures (i.e., unit labeling conventions, mail drop points)
- Address list maintenance using supplemental data sources—comparison of Master Address File with updates from sources not currently used by the Census Bureau, possibly including updates from the Bureau's Demographic Area Address Listing and American Community Survey operations as well as aerial imaging
- Effectiveness of unduplication operations—in preliminary design, may include specific routing of some long-distance duplicates (cases where name and date of birth match but are not geographically proximate) through the Coverage Follow-Up (CFU) operation
- Coverage of Group Quarters Population—based on use of an Alternative Group Quarters Report, as described in Section B–2.c
- Alternative questions in CFU interview—use of questions at the end of CFU interview if responses conflict with answers to the undercount and overcount coverage probe questions on the census forms
- *Other proposed topics:* Administrative records for studying coverage problems; ethnographic studies

**Coverage Measurement**

- *Prospective Topics:* Coverage of group quarters population; reverse record check; quality of Census Coverage Measurement (CCM) data collection and processing; comparison of operations history with CCM results; use of administrative records to augment CCM field work; comparison of household measurement in the census and the CCM survey
- Planned to be completed by fall 2012

**Field Operations**

- Automation for address canvassing—in preliminary design, said to involve comparison of cost and progress data for 2010 address canvassing with similar operations in 2000 as well as summaries of field personnel reactions to use of handheld computers (including Help Desk calls)
- Geographic positioning system (GPS) technology for quality control—in preliminary design, said to involve identification of unusual clusters of GPS coordinates (compared with manually drawn map spots) and use of imagery to examine veracity of clusters

**Language Program**

- Observation of enumerator interactions with non-English-speaking households

**Questionnaire Content**

- Behavior coding of enumerator interviews—analysis of audio tapes of field and telephone center interviews by survey methodologists to study how well interviewers ask, and how respondents answer, questions

*(continued)*

---

**Box B-2** (continued)

- *Prospective Topic:* Comparison of 2010 census and American Community Survey data

**Marketing and Publicity**

- Evaluation of integrated communication program—evaluation to be conducted by National Opinion Research Council, making use of three surveys (including pre- and post-census) study awareness of census outreach

**Privacy and Confidentiality**

- *Prospective Topic:* Public concerns about privacy and confidentiality (possibly linked with communication program evaluation under "Marketing and Publicity")

SOURCES: Presentations and materials shared by the Census Bureau with the panel, particularly Jackson (2008) and Reichert (2009).

---

### B–2.a Alternative Questionnaire Experiment Reinterview

The principal focus of the AQE is differences in response to revised forms of the census questions on race and Hispanic origin; the Census Bureau plans to conduct a reinterview study with about 4,000 households from each of the 15 race and Hispanic-origin treatment groups in the AQE (as well as from the two control groups) to assess response bias. Plans call for the reinterviews to be conducted exclusively by telephone, so the availability of phone numbers (either provided by respondents on the form or found through directory look-ups) will dictate final sample sizes. Because the interest is in response bias, reinterviews will be conducted with the household members who completed the mail census form whenever possible (accepting proxy responses only as a last resort). The reinterview walks through the other questions on the short-form-only census questionnaire but is meant to be more probing with regard to the race and ethnicity questions (i.e., seeking yes/no verification for every major category and subcategory); it is also meant to be more conversational by including an open-ended question on race and origin perceptions ("I'd like you to think about what you *usually* say when asked about your race and origin. . . . Keeping in mind that you can say more than one, what do you usually say when asked about your race and origin?").

### B–2.b Content Reinterview

On a much smaller basis than the AQE reinterview evaluation, the Census Bureau plans to conduct a general content reinterview study similar to those done in previous censuses. Like the AQE, plans call for the Content Reinterview to be performed by telephone, with an estimated sample size of 10,000 interview cases (drawn "proportionally across mail return questionnaires, update/enumerate interviews, nonresponse followup interviews, etc.") in the United States and 860 in Puerto Rico.

### B–2.c  Alternative Group Quarters Questionnaire

The Census Bureau's planned Alternative Group Quarters Questionnaire experiment builds from experience in the 2000 census, but it also portends to repeat a major lost opportunity from that census.

In 2000, every Individual Census Report questionnaire filled by residents of group quarters (college housing, correctional facilities, health care facilities, etc.) included the question: "What is the address of the place where you live or stay MOST OF THE TIME?" (An instruction at the end of the first page of the form was intended to route respondents who live or stay at the group quarters location "most of the time" past this second address question and to the end of the questionnaire; still, the query for the second address dominated the second page of the form.) The same format (with slightly revised wording to fit the circumstances) held for the Military Census Reports and Shipboard Census Reports used to enumerate on-base military personnel and shipboard personnel. Although this "usual home elsewhere" (UHE) query was included on all group quarters questionnaires, the Census Bureau's residence rules for the 2000 census considered UHE information from only certain group quarters types to be valid: valid for military personnel and such small group quarters as temporary worker camps, carnival grounds, and monasteries and convents, but invalid for the most sizable of group quarters populations (e.g., the aforementioned college students, prisoners, or persons in nursing homes).

Like the 2000 census, the Individual Census Report for the 2010 census (Form D-20, filed with OMB in ICR 200808-0607-003) asks about a usual home elsewhere regardless of group quarters type: if a respondent answers "no" to the question "Do you live or stay in this facility MOST OF THE TIME?" he or she is asked "What is the full address of the place where you live or stay MOST OF THE TIME?" The Bureau's planned Alternative Group Quarters Questionnaire takes a different approach by asking for an "any residence elsewhere," regardless of the respondent's answer to the question about living or staying at the group quarters facility most of the time. The experimental question asks: "BESIDES THIS FACILITY, what is the full address of a place where you sometimes live or stay?" No followup question as to how frequently the person lives or stays at this alternate address is asked.

As of its initial March 2009 filing with OMB, the Bureau planned to use this alternative group quarters questionnaire in only 60,000 cases, administering the experimental questions to whole group quarters facilities rather than mixing normal and experimental questionnaires at individual sites. However, in a July 2009 updated filing, the Census Bureau indicated that the sample size had been increased to 125,000, and that this sample would be conducted in only three local census offices (selected in May 2009 "based on demographics and 2010 geography").

The National Research Council (2006) recommended that an "any residence elsewhere" question—similar to this experiment's version but also including a follow-up on frequency of time spent at that address—should be asked of all group quarters residents in 2010 and of a large test sample of non-group-quarters, household respondents. Although this experiment goes a small part toward that recommendation, what is unknown at this time is whether the Bureau will make use of the alternative address information gathered on the standard group quarters form. In 2000, in principle, forms from non-UHE-eligible group quarters types should have been handled separately from those for which UHE was permissible. However, in practice, this prefiltering was not done, and all 2.9 million group quarters forms (including captured other-address responses) went through an initial geocoding operation when only 659,000 should have been eligible. Only when this extra work was done—and the resulting slow-down in other processing noticed—were the UHE-eligible and non-UHE-eligible groups separated, with the other-address information for the non-UHE-eligible group quarters types being discarded. The National Research Council (2006:230) described this situation as a "highly regrettable lost research opportunity"—throwing out what could have been a "trove of information on the nature of potential census duplicates."

### B–2.d Interactive Voice Response Customer Satisfaction Survey

The Census Bureau plans to use interactive voice response (IVR) technology in its Telephone Questionnaire Assistance (TQA) program in 2010. Respondents seeking clarification or information (e.g., requesting a foreign-language questionnaire) can call a number in the census mailing package, and an automated computer system will administer help based on spoken word commands from the respondent. As one of the formal CPEX evaluations, the Census Bureau plans to conduct a Customer Satisfaction Survey with an approximate 1 percent sample of persons calling the TQA lines to assess satisfaction with the IVR interface. The current plan is for this survey to be done at the end of a TQA call; the IVR system will say:

> If that's all the information you needed, please hold for our Customer Satisfaction Survey. Otherwise, to hear the topic information again say "repeat that" or for help on another general question say "Census information."

About two-thirds of the total estimated sample size (665,000 callers) will then be administered a five-item satisfaction survey via IVR, while the other third will be transferred to a customer service representative to receive a seven-question survey. Perhaps recognizing that customers generally unsatisfied with speaking to the IVR system will not be inclined to go through a follow-up survey on IVR (particularly one administered by IVR)—and possibly accounting for people whose inquiries lead them to be diverted out of the IVR to a human operator—the Census Bureau currently projects a 7.6 percent response rate and an estimated 5,016 total respondents.

## B–3   ASSESSMENTS

Box B-3 lists what is currently known about the content of the assessments portion of the CPEX program, in which "assessment" refers to operational histories and descriptions akin to what were generally labeled "evaluations" in 2000. Aside from these major headings, we have no more detailed information about research plans for specific assessments.

---

**Box B-3** Assessments in the 2010 CPEX

**Assessment Studies, Grouped by Name of Designated Staff Teams**

- *Content and Forms Design Integrated Product Team*—Content and forms design; item nonresponse and imputation rates
- *Field Infrastructure–Administrative Operational Integration Team*—Field office administration and payroll; recruiting and hiring field staff; regional census center/local census office leasing, space, equipment, and information technology; Census Hiring and Employment Check (CHEC) system; Decennial Applicant, Personnel, and Payroll System (DAPPS)
- *Field Infrastructure–Field Activities Operational Integration Team*—Fingerprinting; kits
- *Language Program Integrated Product Team*—Language fulfillment program; bilingual questionnaire
- *Integrated Communication Operational Integration Team*—Integrated Communications Program
- *Geographic Programs Operational Integration Team*—Type of enumeration area delineation; Local Update of Census Addresses (LUCA) participation, review, and processing; LUCA feedback, appeals, and processing; service-based enumeration, group homes, and carnival locations address list update; new construction
- *Address List Development Operational Integration Team*—Non-ID processing; field verification; update/leave; address canvassing; group quarters validation
- *Universe Control and Management System Integrated System Team*—Universe control and management system
- *Forms Printing and Distribution Integrated Product Team*—Forms and printing; removing undeliverable as addressed (UAAs) from the replacement questionnaire workload/postal tracking for optimal replacement questionnaire cutoff; mail response and return rates
- *Housing Unit Enumeration Operational Integration Team*—Enumeration of transitory locations; update/enumerate; remote Alaska; nonresponse follow-up and vacant/delete check (includes removing late mail returns); Coverage Follow-Up; Remote Update/Enumerate; Be Counted/Questionnaire Assistance Centers (cosponsored with Integrated Communication Team)
- *Group Quarters Enumeration Operational Integration Team*—Military enumeration; group quarters enumeration, including advance visit; service-based enumeration
- *Island Areas Operational Integration Team*—Island Areas
- *Telephone Questionnaire Assistance and Fulfillment Operational Integration Team*—Telephone Questionnaire Assistance program
- *Decennial Response Integration System*—Data capture and integration
- *Response Processing System Integrated System Team*—Response processing system; Census Unedited File; Census Edited File
- *Count Review Integrated Product Team*—Count review
- *Archiving Integrated Product Team*—Archiving
- *Census Coverage Measurement Operational Integration Team*—Initial housing unit (listing, matching, and follow-up); person interview; final housing unit; sample design; person matching and follow-up
- *Demographic Analysis Research Operational Integration Team*—Demographic analysis
- *Cost and Progress*—Cost and Progress System
- *2010 Planning and Coordination Staff*—Program management processes (change control, schedule, risks, issues, and requirements)

SOURCE: List of assessments shared by U.S. Census Bureau with the panel, May 15, 2009.

Part II

# Interim Report: *Experimentation and Evaluation in the 2010 Census* (December 7, 2007)

# Executive Summary

In connection with every recent decennial census, the U.S. Census Bureau has carried out experiments and evaluations. A census "experiment" usually involves field data collection during the census in which alternatives to current census processes are assessed for a subset of the population. An "evaluation" is usually a post hoc analysis of data collected as part of the decennial census processing to determine whether individual steps in the census operated as expected. The Census Bureau program for evaluations and experiments for the 2010 decennial census is referred to as the 2010 CPEX Program.

CPEX, like its predecessor programs, has enormous potential to help improve the next census, which is the federal government's single most important, and most costly, data collection activity. A well-planned and well-executed CPEX is a sound investment to ensure that the 2020 census is as cost-effective as possible.

The Census Bureau is now determining the topics for experiments during the 2010 census. The specific designs of the experiments have to be final by summer 2008 to meet the planning needs for the census. Because the data needed to support census evaluations are typically output files from the census itself, the exact structure of individual evaluations is not yet as time-sensitive as the experiments. However, some early planning for evaluations is crucial so that the necessary data extracts can be prepared and retained. This is especially true because much of the data collection in 2010 will be carried out by contractors, and so data retention requirements need to be arranged with contractors as early as possible.

The Panel on the Design of the 2010 Census Program of Evaluations and Experiments has been broadly charged to review proposed topics for evaluations and experiments and recommend priorities for them for the 2010 census, to consider what can be learned from the 2010 testing cycle to better plan for the 2020 census, and to assess the Census Bureau's overall continuing research program for the nation's decennial censuses.

The primary purpose of this interim report is to help reduce the possible subjects for census experimentation from an initial list of 52 research topics compiled by the Census Bureau to perhaps 6, which is consistent with the size of the experimentation program in 2000. This interim report also offers broad advice on plans for evaluations of the 2010 census. The panel expects to provide fuller details of individual experiments and evaluations in its subsequent reports.

## CENSUS EXPERIMENTS

The panel identified three priority experiments for inclusion in the 2010 census to assist 2020 census planning (in one instance, there might be several related experiments): an experiment on the use of the Internet for data collection; an experiment on the use of administrative records for various census purposes; and an experiment (or set of experiments) on features of the census questionnaire.

One important opportunity for improving census quality and possibly reducing census costs in 2020 is the use of the Internet as a means of enumeration. Although Internet response was permitted (but not advertised) in the 2000 census, the Census Bureau has elected not to allow online response in 2010. The panel does not second-guess that decision, but we think that it is essential to have a full and rigorous test of Internet methodologies in the 2010 CPEX. Internet response provides important advantages for data collection, including alternate ways of presenting residence rules and concepts, increased facility for the presentation of questionnaires in foreign languages, and real-time editing. It also has the feature of immediate transmission of data, which has important benefits regarding minimizing the overlap of census data collection operations.

> *Recommendation 1:* **The Census Bureau should include, in the 2010 census, a test of Internet data collection as an alternative means of enumeration. Such a test should investigate means of facilitating Internet response and should measure the impact on data quality, the expeditiousness of response, and the impact on the use of foreign language forms.**

Another important opportunity for reducing costs and improving data quality is the use of administrative records. These are data collected as a by-product of the management of federal, state, and local governmental programs, such as birth and death records, building permit records, and welfare program records. In 2000, administrative records were the subject of an experiment intended to study their use as a complementary type of enumeration (that is, whether person counts for some geographic areas derived from records were consistent with census returns). However, administrative records could be used more broadly to assist a number of census tasks, including such uses as (1) to improve the Master Address File, (2) as an alternative to last-resort proxy response, (3) as an alternative to item and unit imputation, (4) to resolve duplicate search, (5) to validate edit protocols, (6) for coverage measurement and coverage evaluation, (7) for coverage improvement, and (8) to help target households for various purposes. It is important for the Census Bureau to determine, starting now, which of these various potential uses of administrative records would or would not be effective for use in 2020.

> *Recommendation 2:* The Census Bureau should develop an experiment (or evaluation) that assesses the utility of administrative records for assistance in specific census component processes—for example, for improvement of the Master Address File, for nonresponse follow-up, for assessment of duplicate status, and for coverage improvement. In addition, either as an experiment or through evaluations, the Census Bureau should collect sufficient data to support assessment of the degree to which targeting various census processes, using administrative records, could reduce census costs or improve census quality.

Finally, given the crucial importance of the census questionnaire as a driver of census data quality, especially with regard to the nation's data on race and ethnicity, and to correctly locate each person at the proper census residence, the Census Bureau should conduct either a large experiment or several smaller experiments on the content and method of presentation of the census questionnaire.

> *Recommendation 3:* The Census Bureau should include one or more alternate questionnaire experiments during the 2010 census to examine:
>
> - the representation of questions on race and ethnicity on the census questionnaire, particularly asking about race and Hispanic origin as a single question;
> - the representation of residence rules and concepts on the census questionnaire; and
> - the usefulness of including new or improved questions or other information on the questionnaire with regard to (1) coverage probes, (2) the motivation of census questions, (3) the request of information on usual home elsewhere on group quarters questionnaires, and (4) deadline messaging and mailing dates for questionnaires.

In such experiments, both the 2000 and the 2010 census questionnaires should be included in the assessments to serve as controls. The Census Bureau should explore the possibility of joining the recommended experiments listed above into a single experiment, through use of fractional factorial experimental designs.

## CENSUS EVALUATIONS

It is important that sufficient data be retained to enable postcensus evaluations of the processes used to update the Master Address File from census to census. The success of a mailout-mailback census is most dependent on the quality of its address list, and therefore understanding the contribution of the various processes used to update the address list, especially Local Update of Census Addresses (LUCA) and address canvassing, is crucially important. In addition, given the expense of address canvassing in all blocks, it is important to be able to ascertain the extent to which canvassing can be targeted to blocks that are likely to have changes. Both administrative records, especially building permit data, and commercial mailing lists may have value in assisting in the targeting of blocks for canvassing.

> *Recommendation 4:* The Census Bureau should design its Master Address File so that the complete operational history—when list-building operations have added, deleted, modified, or simply replicated a particular address record—can be reconstructed. This information will support a comprehensive evaluation of the Local Update of Census Addresses and address canvassing. In addition, sufficient information should be retained, including relevant information from administrative records and the American Community Survey, to support evaluations of methods for targeting blocks that may not benefit from block canvassing. Finally, efforts should be made to obtain addresses from commercial mailing lists to determine whether they also might be able to reduce the need for block canvassing.

More broadly, a master trace sample database could be used to address a substantial number of questions about the functioning of the 2010 census. Such a database would necessitate the retention of the entire census processing history (including the coverage measurement processes) of all addresses for a selected sample of areas, structured in a way to facilitate analysis. For example, such a database would help determine what percentage of census omissions are in partially enumerated households, or it could assess the benefits of the coverage follow-up interview. The panel therefore recommends that the process for creating such a database be initiated.

*Recommendation 5:* The Census Bureau should initiate efforts now for planning the general design of a master trace sample database and should plan for retention of the necessary information to support its creation.

Also, evaluations should be carried out on the feasibility of coverage measurement through use of a reverse record check based on the American Community Survey. The reverse record check is an alternative method for estimating the completeness of census coverage of the population, which may have advantages over the methods of dual-systems estimation and demographic analysis that have been used for this purpose to date.

*Recommendation 6:* The Census Bureau, through the use of an evaluation of the 2010 census (or an experiment in the 2010 census) should determine the extent to which the American Community Survey could be used as a means for evaluating the coverage of the decennial census through use of a reverse record check.

Finally, the Census Bureau has no program for assessing the rate of omissions of residents of group quarters in the 2010 census, nor can it assess the rate of placement of group quarters in the wrong census geography. The Census Bureau should therefore take the first steps toward remedying this by collecting sufficient information in 2010 to evaluate ideas on how to include this capability in the 2020 census coverage measurement program.

*Recommendation 7:* The Census Bureau should collect sufficient data in 2010 to support the evaluation of potential methods for assessing the omission rate of group quarters residents and the rate of locating group quarters in the wrong census geography. This is a step toward the goal of improving the accuracy of group quarters data.

## OVERALL CENSUS RESEARCH PROGRAM

It appears that basic census research is not receiving the priority and support needed to best guide census redesign. For example, tests on some topics have been unnecessarily repeated, and previous research has sometimes been ignored in designing newer tests. Also, some topics, by their nature, require a relatively long time to understand and therefore need to be separated from the decennial census operational cycle. The lack of priority of research can also be seen in that the results of the 2006 test census tests were not all completed in time for the design of the 2008 census dress rehearsal. Research continuity is important not only to reduce redundancy and to ensure that findings are known and utilized, but also because there are a number of issues that come up repeatedly over many censuses that are inherently complex and therefore benefit from testing in a variety of circumstances in an organized way, as unaffected as possible by the census cycle. Finally, given the fielding of the American Community Survey, there is now a real opportunity for research on census and survey methodology to be more continuous.

> *Recommendation 8:* The Census Bureau should support a dedicated research program in census and survey methodology, whose work is relatively unaffected by the cycle of the decennial census. In that way, a body of research findings can be generated that will be relevant to more than one census and to other household surveys.

## THE 2010 CENSUS DESIGN

In carrying out our charge to advise on the development of plans for experimentation and evaluation for the 2010 census, and more generally to review the full program of research and testing for improving census methodology, three issues arose that relate to the 2010 census design itself and, consequently, its evaluation. While the panel is aware that most aspects of the 2010 census design have already been decided and cannot be easily changed given time constraints, there remains the possibility that some of the following recommendations may still be able to be acted on prior to 2010.

The first issue is the possibility of the introduction of errors into the data collection transmissions by the handheld computing devices that will be used to follow up households that do not return a mail questionnaire. The second issue is the possibility of interoperability problems in the various software systems constituting the management information system for the 2010 census. The third issue is the role of telephone questionnaire assistance in 2010.

> *Recommendation 9:* The Census Bureau should use dual-recording systems, quantitative validation metrics, dedicated processing systems, periodic system checkpoints, strict control over handheld devices, and related techniques to ensure and then verify the accuracy of the data collected from handheld computing devices.

> *Recommendation 10:* The Census Bureau should provide for a check to ensure that the subsystems of the management information system used in 2010 have no interoperability problems.

> *Recommendation 11:* The Census Bureau should strongly consider, for the 2010 census, explicit encouragement of the collection of all data on the census questionnaire for people using Tele-

phone Questionnaire Assistance. In addition, the Census Bureau should collect sufficient information to estimate the percentage of callers to Telephone Questionnaire Assistance who did not ultimately send back their census questionnaires. This would provide an estimate of the additional costs of nonresponse follow-up due to the failure to collect the entire census questionnaire for those cases. The Census Bureau should also consider carrying out an experiment whereby a sample of callers to Telephone Questionnaire Assistance are asked whether they would mind providing their full information to better estimate the additional resources required as a result of expanding Telephone Questionnaire Assistance in this way.

# – 1 –

# Introduction

The Census of Population and Housing is carried out in the United States every 10 years, and the next census is scheduled to begin its mailout-mailback operations in March 2010. For at least the past 50 years, each decennial census has been accompanied by a research program of evaluation or experimentation. The Census Bureau typically refers to a census "experiment" as a study involving field data collection—typically carried out simultaneously with the decennial census itself—in which alternatives to census processes currently in use are assessed for a subset of the population. By comparison, census "evaluations" are usually post hoc analyses of data collected as part of the decennial census process to determine whether individual steps in the census operated as expected. Collectively, census experiments and evaluations are designed to inform the Census Bureau as to the quality of the processes and results of the census, as well as to help plan for modifications and innovations that will improve the (cost) efficiency and accuracy of the next census. The Census Bureau is currently developing a set of evaluations and experiments to accompany the 2010 census, which the Bureau refers to as the 2010 Census Program for Evaluations and Experiments or CPEX.

These two activities of the more general census research program are concentrated during the conduct of the census itself, but census-related research activities continue throughout the decade. Traditionally, the Census Bureau's intercensal research has been focused on a series of census tests, some of which are better described as "test censuses" because they are conducted in specific geographic areas and can include fieldwork (e.g., in-person follow-up for nonresponse) as well as contact through the mail or other means. The sequence of tests usually culminates in a dress rehearsal two years prior to the decennial census. In addition to the test censuses, the Census Bureau has also conducted some smaller scale experimental data collections during the intercensal period.

## 1–A   CHARGE TO THE PANEL

As it began to design its CPEX program for 2010, the Census Bureau requested that the Committee on National Statistics of the National Academies convene the Panel on the Design of the 2010 Census Program of Evaluations and Experiments. The panel's charge is to:

> ... consider priorities for evaluation and experimentation in the 2010 census. [The panel] will also consider the design and documentation of the Master Address File and operational databases to facilitate research and evaluation, the design of experiments to embed in the 2010 census, the design of evaluations of the 2010 census processes, and what can be learned from the pre-2010 testing that was conducted in 2003–2006 to enhance the testing to be conducted in 2012–2016 to support census planning for 2020. Topic areas for research, evaluation, and testing that would come within the panel's scope include questionnaire design, address updating, nonresponse follow-up, coverage follow-up, unduplication of housing units and residents, editing and imputation procedures, and other census operations. Evaluations of data quality would also be within scope. . . .

More succinctly, the Census Bureau requests that the panel:

- Review proposed topics for evaluations and experiments;
- Assess the completeness and relevance of the proposed topics for evaluation and experimentation;
- Suggest additional research topics and questions;
- Recommend priorities;
- Review and comment on methods for conducting evaluations and experiments; and
- Consider what can be learned from the 2010 testing cycle to better plan research for 2020.

The panel is charged with evaluating the 2010 census research program, primarily in setting the stage for the 2020 census. As the first task, the panel was asked to review an initial list of research topics compiled by the Census Bureau, with an eye toward identifying priorities for specific experiments and evaluations in 2010. This first interim report by the panel uses the Bureau's initial suggestions for consideration as a basis for commentary on the overall shape of the research program surrounding the 2010 census and leading up to the 2020 census. It is specifically the goal of this report to suggest priorities for the experiments to be conducted in line with the 2010 census because they are the most time-sensitive. To some observers, a two-year time span between now and the fielding of the 2010 census may seem like a long time; in the context of planning an effort as complex as the decennial census, however, it is actually quite fleeting. Experimental treatments must be specified, questionnaires must be tested and approved, and systems must be developed and integrated with standard census processes—all at the same time that the Bureau is engaged in an extensive dress rehearsal and final preparations for what has long been the federal government's largest and most complex non-military operation. Accordingly, the Census Bureau plans to identify topics for census experiments to be finalized by winter 2007 and to have more detailed plans in place in summer 2008; this report is an early step in that effort.

Although this report is primarily about priorities for experiments, we also discuss the evaluation component of the CPEX. This is because even the basic possibilities for specific evaluations depend critically on the data that are collected during the conduct of the census itself. Hence, we offer comments about the need to finalize plans for 2010 data collection—whether in house by the Census Bureau or through its technical contractors—in order to facilitate a rich and useful evaluation program.

We will continue to study the CPEX program over the next few years, and we expect to issue at least one more report; these subsequent reports will respond to the Bureau's evolving development of the CPEX plan as well as provide more detailed guidance for the conduct of specific evaluations and experiments.

## 1–B  BACKGROUND: EXPERIMENTS AND EVALUATIONS IN THE 2000 CENSUS

As context for the discussion that follows and to get a sense of the scope of CPEX, it is useful to briefly review the experiments and evaluations of the previous census. The results of the full Census 2000 Testing, Experimentation, and Evaluation Program are summarized by Abramson (2004).

### 1–B.1  Experiments

The Census Bureau carried out five experiments in conjunction with the 2000 census. Several ethnographic studies were also conducted during the 2000 census; about half of these were considered to be part of the formal evaluation program, whereas the others were designated as a sixth experiment.

- **Census 2000 Alternative Questionnaire Experiment (AQE2000):** AQE2000 comprised three experiments for households in the mailout-mailback universe of the 2000 census. The *skip instruction experiment* examined the effectiveness of different methods for guiding respondents through an alternative long-form questionnaire with skip patterns. The *residence instructions experiment* tested various methods (format, presentation, and wording of instructions) for representing the decennial census residence rules on the questionnaire. The hope was to improve within-household coverage by modifying the roster instructions. Finally, the *race and Hispanic origin experiment* compared the 1990 race and Hispanic origin questions with the questions on the Census 2000 short form, specifically assessing the effect of permitting the reporting of more than one race and reversing the sequence of the race and Hispanic origin items. This experiment is summarized by Martin et al. (2003).

- **Administrative Records Census 2000 Experiment (AREX 2000):** AREX 2000 was designed to assess the value of administrative records data in conducting an administrative records census. As a by-product, it also provided useful information as to the value of administrative records in carrying out or assisting in various applications in support of conventional decennial census processes. AREX 2000 used administrative records to provide information on household counts, date of birth, race, Hispanic origin, and sex, linked to a corresponding block code.

  The test was carried out in five counties in two sites, Baltimore City and Baltimore County, Maryland, and Douglas, El Paso, and Jefferson counties in Colorado, with approximately 1 million housing units and a population of approximately 2 million. The population coverage for the more thorough of the schemes tested was between 96 and 102 percent relative to the Census 2000 counts for the five test site counties. However, the AREX 2000 and the census counted the same number of people in a housing unit only 51.1 percent of the time. They differed by at most one person only 79.4 percent of the time. The differences between the administrative records–based counts and the census counts were primarily attributed to errors in address linkage and typical deficiencies in administrative records (missed children, lack of representation of special populations, and deficiencies resulting from the time gap between the administrative records extracts and Census Day). Another important finding was that administrative records are not currently a good source of data for race and Hispanic origin, and the models used to impute race and Hispanic origin were not sufficient to correct the deficiencies in the data. The experiment is summarized by Bye and Judson (2004).

- **Social Security Number, Privacy Attitudes, and Notification Experi-**

ment (SPAN): This experiment assessed the public's attitudes regarding the census and its uses, trust and privacy issues, the Census Bureau's confidentiality practices, possible data sharing across federal agencies, and the willingness of individuals to provide their Social Security number on the decennial census questionnaire. In addition, the public's attitude toward the use of administrative records in taking the census was also assessed. The experiment is described in detail by Larwood and Trentham (2004).

- **Response Mode and Incentive Experiment (RMIE):** The RMIE investigated the impact of three computer-assisted data collection techniques: computer-assisted telephone interviewing (CATI), the Internet, and interactive voice response, on the response rate and quality of the data collected. The households in six panels were given the choice of providing their data via the usual paper forms or by one of these alternate modes. Half of the panels were offered an incentive—a telephone calling card good for 30 minutes of calls—for using the alternate response mode. In addition, the experiment included a nonresponse component designed to assess the effects on response of an incentive to use alternative response mode options among a sample of census households that failed to return their census forms by April 26, 2000. This was to test the effect of these factors on a group representing those who would be difficult to enumerate. A final component of this experiment involved interviewing households assigned to the Internet mode who opted to complete the traditional paper census form to determine why these households did not use the Internet. One of the findings was that the Internet provided relatively high-quality data. However, among respondents who were aware of the Internet option, 35 percent reported that they believed the paper census form would be easier to complete. Other reasons for not using the Internet include no access to a computer, concerns about privacy, "forgot the Internet was an option," and insufficient knowledge of the Internet. The incentive did not increase response but instead redirected response to the alternate modes. The CATI option seemed to be preferred over the other two alternate modes. Caspar (2004) summarizes the experiment's results.

- **Census 2000 Supplementary Survey (C2SS):** By 1999, the basic notion that the new American Community Survey (ACS) would take the role of the traditional census long- form sample had been established (this is discussed in more detail in the next section). ACS testing had grown to include fielding in about 30 test sites (counties), with full-scale implementation planned for the 2000-2010 intercensal period. Hence, the Census Bureau was interested in some assessment of the operational feasibility of conducting a large-scale ACS at the same time as a decennial census. Formally an experiment in the 2000 census program, the C2SS escalated ACS data collection to include more than one-third of all counties in the United States; this step-up in collection—while well short of full-scale implementation—offered a chance to compare ACS estimates with those from the 2000 census. Operational feasibility was defined as the C2SS tasks being executed on time and within budget with the collected data meeting basic quality standards. No concerns about the operational feasibility of taking the ACS in 2010 were found. Griffin and Obenski (2001) wrote a summary report on the operational feasibility of the ACS, based on the C2SS.

• **Ethnographic Studies:** Three studies were included in this experiment. One study examined the representation of and responses from complex households in the decennial census through ethnographic studies of six race/ethnic groups (Schwede, 2003). A second study examined shared attitudes among those individuals following the "baby boomers," i.e., those born between 1965 and 1975, about civic engagement and community involvement, government in general, and decennial census participation in particular (Crowley, 2003). A third study examined factors that respondents considered when they were asked to provide information about themselves in a variety of modes (Gerber, 2003). This research suggested that the following factors may contribute to decennial noncompliance and undercoverage errors: (1) noncitizenship status or unstable immigration status, (2) respondents not knowing about the decennial census, and (3) increased levels of distrust among respondents toward the government.

## 1–B.2   Evaluations

The Census Bureau initially planned to conduct 149 evaluation studies to assess the quality of 2000 census operations. Due to various resource constraints, as well as the overlap of some of the studies with assessments needed to evaluate the quality of the 2000 estimates of net undercoverage, 91 studies were completed. These evaluations were summarized in various topic reports, the subjects of which are listed in Table 1-1.

## 1–C   POST HOC ASSESSMENT OF THE 2000 EXPERIMENTS AND EVALUATIONS

We have described six experiments that were embedded in the 2000 census. We can now look back at these experiments to see the extent to which they were able to play a role in impacting the design of the 2010 census. In doing that we hope to learn how to improve the selection of experiments in

**Table 1-1** Topic Headings, 2010 CPEX Research Proposals and 2000 Census Evaluation Program

| 2010 CPEX Proposals | 2000 Census Evaluation Topic Reports |
|---|---|
| Content | Content and data quality |
| Coverage improvement | Coverage improvement |
| Address list development | Address list development |
| Administrative records | *AREX2000 experiment*[a] |
| Coverage follow-up | *Partial: Coverage improvement* |
| Residency rules/question development | *AQE2000 experiment* |
| Be Counted | *Partial: Response rates and behavior analysis* |
| General | — |
| Coverage measurement | Coverage measurement |
| Field activities | |
| Automation | *Partial: Automation of census 2000 processes* |
| Training | — |
| Quality control | *Partial: Content and data quality* |
| Language | *Partial: Response rates and behavior analysis* |
| Marketing/publicity/advertising/partnerships | Partnership and marketing program |
| Mode effects | — |
| Privacy | Privacy research in census 2000[b] |
| Race and Hispanic origin | Race and ethnicity |
| Self-response options | — |
| Special places and group quarters | Special places and group quarters |
| — | Automation of census 2000 processes |
| — | Data capture |
| — | Data collection |
| — | Data processing |
| — | Ethnographic studies[c] |
| — | Puerto Rico |
| — | Response rates and behavior analysis |

NOTE: The italics in the entries indicate deviations from the column heading, "2000 Census Evaluation Topic Reports." Some of the entries were not topic reports but were experiments. Also, some of the operations were part of the 2000 Coverage Improvement report.

[a] Described as partial match because the CPEX proposals under automation are oriented principally at one component (handheld computers).

[b] Privacy was also touched on by the Social Security Number, Privacy Attitudes, and Notification (SPAN) experiment.

[c] The 2000 census included several ethnographic studies; administratively, about half were considered part of the experiments while others were formally designated as evaluations (and were the subject of a topic report).

the 2010 census, looking toward the design of the 2020 census. Before continuing, it is important to note that the very basic design of the 2010 census was determined before these 2000 census experiments had been carried out. Therefore, at a fundamental level, the 2000 census experiments were always limited in their impact on key aspects of the basic design of the next census.

On the one hand, with the benefit of hindsight, the choice of the general subject matter for these six experiments can be viewed as relatively successful, since many of the basic issues identified for experimentation were relevant to the design of the 2010 census. The utility of information from administrative records for census purposes, the advantages and disadvantages of Internet data collection, various aspects of census questionnaire design, and the operational feasibility of the American Community Survey being carried out during a decennial census were issues for which additional information was needed to finalize the 2010 design.

On the other hand, the details of these studies also indicate that they could have played a more integral role in the specification of the design for the 2010 census if they had been modified in relatively modest ways. For example, as a test of residence instructions, AQE2000 varied many factors simultaneously so that individual design effects were difficult to separate out. Also, the test of long-form routing instructions was largely irrelevant to a short-form-only census. AREX 2000 focused on the use of administrative records to serve in place of the current census enumeration, whereas examination of the use of administrative records to help with specific operations, such as for targeted improvements in the Master Address File, to assist in late nonresponse follow-up, or to assist with coverage measurement, would have been more useful. The response mode and incentive experiment examined the use of incentives to increase use of the Internet as a mode of response, but they did not examine other ways to potentially facilitate and improve Internet usage. Finally, the Social Security Number, Privacy, and Notification Experiment did not have any direct bearing on the 2010 design.

It bears repeating that it is an enormous challenge to anticipate what issues will be prominent in census designs for a census that will not be finalized for at least eight years after the census experiments themselves need to be finalized. Since one goal of this panel study is to help the Census Bureau select useful experiments for the 2010 census, our hope is that, when looking back in 2017, the 2010 census experiments will be seen as very useful in helping to select an effective design for the 2020 census.

With respect to the 2000 evaluations, the National Research Council report *The 2000 Census: Counting Under Adversity* provided an assessment of the utility of these studies, with which we are in agreement. The study group found (National Research Council, 2004a:331–332):

> Many of the completed evaluations are accounting-type documents rather than full-fledged evaluations. They provide authoritative infor-

mation on such aspects as number of mail returns by day, complete-count item nonresponse and imputation rates by type of form and data collection mode, and enumerations completed in various types of special operations. . . . This information is valuable but limited. Many reports have no analysis as such, other than simple one-way and two-way tabulations. . . . Almost no reports provide tables or other analyses that look at operations and data quality for geographic areas. . . . 2010 planners need analysis that is explicitly designed to answer important questions for research and testing to improve the 2010 census. . . . Imaginative data analysis [techniques] could yield important findings as well as facilitate effective presentation of results.

## 1–D   OVERVIEW OF THE 2010 CENSUS

While the 2000 census was still under way, the Census Bureau began to develop a framework for the 2010 census. Originally likened to a three-legged stool, this framework was predicated on three major initiatives:

- The traditional long-form sample—in which roughly one-sixth of census respondents would receive a detailed questionnaire covering social, economic, and demographic characteristics—would be replaced by a continuing household survey throughout the decade, the American Community Survey, thus freeing the 2010 census to be a short-form-only enumeration;

- Improvements would be made to the Census Bureau's Master Address File (MAF) and its associated geographic database (the Topologically Integrated Geographic Encoding and Referencing System, or TIGER, database) in order to save field time and costs; and

- A program of early, integrated planning would be implemented in order to forestall an end-of-decade crunch in finalizing a design for the 2010 census.

*Reengineering the 2010 Census: Risks and Challenges* reviews the early development of the 2010 census plan, noting an immediate adjunct to the basic three-legged plan: the incorporation of new technology in the census process (National Research Council, 2004b). Specifically, the 2010 census plan incorporated the view that handheld computers could be used in several census operations in order to reduce field data collection costs and improve data quality. Following a series of decisions not to adjust the counts from the 2000 census for estimated coverage errors, the Census Bureau also established the basic precept that the 2010 census coverage measurement program would be used primarily to support a feedback loop of census improvement rather than for census adjustment.

As the 2010 census plan has developed, major differences between the 2010 plan and its 2000 predecessor—in addition to the broad changes already described—include:

- The use of handheld computers by field enumerators has been focused on three major operations: updating the Master Address File during the address canvassing procedure, conducting nonresponse follow-up interviewing, and implementing a new coverage follow-up (CFU) operation.

- The coverage follow-up interview is a consolidation and substantial expansion of a telephone follow-up operation used in the 2000 census, which was focused on following up households with count discrepancies and households with more than the six maximum residents allowed on the census form. While detailed plans for this follow-up operation are as yet incomplete, it appears that the CFU in 2010 will also follow up households with evidence of having duplicate enumerations, with people viewed as residents who possibly should have been enumerated elsewhere, and with people viewed as nonresidents who may have been incorrectly omitted from the count of that household.

- The Local Update of Census Addresses (LUCA) program, which gives local and tribal governments an opportunity to review and suggest additions to the Master Address File from their areas, has been revised to facilitate participation by local governments and to enhance communication between Census Bureau and local officials.

- Nonrespondents to the initial questionnaire mailing will be sent a replacement questionnaire to improve mail response.

- Households in selected geographic areas will be mailed a bilingual census questionnaire in Spanish and English.

- The census questionnaire will include two "coverage probe" questions to encourage correct responses (and to serve as a trigger to inclusion in the CFU operation).

- The definitions of group quarters—nonhousehold settings like college dormitories, nursing homes, military barracks, and correctional facilities—have been revised.

- Continuing a trend from 2000, the Census Bureau will increasingly rely on outside contractors to carry out several of the processes.

## 1–E  THE CPEX PLANNING DOCUMENT

This, the panel's first interim report, provides a review of the current status of the experimentation and evaluation plans of the Census Bureau heading into the 2010 census. As the major input to the panel's first meeting and our work to date, the Census Bureau provided a list of 52 issues, reprinted as Appendix A, corresponding to component processes of the 2010 census design that were viewed either as potentially capable of improvement or of sufficient concern to warrant a careful assessment of their performance in 2010. The list, divided into the following 11 categories, was provided to us as the set of issues that the Census Bureau judged as possibly benefiting from either experimentation in 2010 or evaluation after the 2010 census has concluded:

- content,
- race and Hispanic origin,
- privacy,
- language,
- self-response options,
- mode effects,
- special places and group quarters,
- marketing/publicity,
- field activities,
- coverage improvement, and
- coverage measurement.

In addition to the description of the topics themselves, the Census Bureau also provided indications as to whether these topics have a high priority, whether they could potentially save substantial funds in the 2020 census, whether results could conclusively measure the effects on census data quality, whether the issue addresses operations that are new since the 2000 census, and whether data will be available to answer the questions posed.

This list of topics was a useful start to the panel's work, but, as discussed more below, it is deficient in some ways, especially since it is not separated into potential experiments or evaluations and does not contain quantitative information on cost or quality implications. Also, such a list of topics needs to be further considered in the context of a general scheme for the 2020 census.

## 1–F   GUIDE TO THE REPORT

The remainder of this report is structured as follows. Chapter 2 provides initial views on the 2010 census experiments. There is a first section on a general approach to the selection of census experiments, which is followed by the panel's recommended priorities for topics for experimentation in 2010. Chapter 3 begins with suggestions for the 2010 census evaluations, which is followed by a general approach to census evaluation, and which concludes with considerations regarding a general approach to census research. Chapter 4 presents additional considerations for the 2010 census itself. It begins with technology concerns for 2010, followed by a discussion of the issue of data retention by census contractors. The chapter concludes with a discussion of the benefits of facilitating census enumeration as part of telephone questionnaire assistance. Appendix A provides the Census Bureau's summaries of suggested research topics for experiments and evaluations in 2010. Appendix B summarizes Internet response options in the 2000 U.S. census and in selected population censuses in other countries. Appendix III presents biographical sketches of panel members and staff.

# – 2 –

# Initial Views on 2010 Census Experiments

## 2–A  A GENERAL APPROACH TO THE SELECTION OF CENSUS EXPERIMENTS

The Census Bureau provided the panel with a list of 52 topics for experimentation or evaluation, categorized into 11 general headings (see Appendix A). In addition to the topics themselves, the Census Bureau provided indications as to (a) whether modification of the relevant census processes have a high priority, (b) whether modification of the relevant census processes could potentially save substantial funds in the 2020 census, (c) whether results of an experiment could conclusively measure the effects on census data quality, (d) whether the issue addresses operations that are new since the 2000 census, and (e) whether data will be available to answer the particular questions posed. The panel found these topics and the associated assessments very helpful in focusing our work. The assessments of these topics, in particular, represent a considerable advance over the processes used to select the evaluations and experiments prior to the 2000 census.

However, we think that the Census Bureau can go further, when preparing for the analogous 2020 CPEX program, by providing a more developed context for evaluating various topics for potential census experiments. It is difficult to develop priorities without some sense of the collection of census designs that are under serious consideration. For example, it was not useful, at least from a decennial census perspective, to test skip patterns for the long form in 2000 given that the likely design in 2010 was a short-form-only census (although it may have been useful in support of the American Community Survey). Similarly, it was not useful to test an administrative records census in the Administrative Records Census 2000 Experiment when that was a remote possibility for the 2010 census. We understand that it will not be possible for the Census Bureau to produce a single proposal for the general design of the next census when it is time to select the experiments and evaluations for the current census, but it should be possible to produce a relatively small number of leading alternative designs that are under consideration. To help define possible designs, fundamental questions like the following might be asked:

- Could the telephone or the Internet be used more broadly as an alternative to mailing back census questionnaires for data collection?

- Could administrative records or other data sources be used to better target various operations?

- Could administrative records be used to augment last-resort or proxy enumeration in the latter stages of nonresponse follow-up?

Having a set of designs that are under consideration helps to direct the experimentation toward resolving important issues that discriminate among the designs.

Although we realize that the following are not readily available, in the future it would also be useful to have, for both the current census processes and, to the extent possible, any alternative approaches: (1) estimates of census costs by component operation (and the recent history of costs)[1] and (2) the potential impact on the quality of the collected data by component operation. The attribution of both coverage and characteristics error, to component operations or current processes, let alone suggested alternatives, on a national level, not to mention for demographic subgroups, would have been very difficult to achieve in past censuses. The planned census coverage measurement program in 2010 is hoping to make progress in assessing and attributing component coverage error to various sources. This is an important development because the Census Bureau could better justify priorities in undertaking various experiments by providing information on the impact on costs and quality of various alternatives. Furthermore, even if estimates of costs and impacts on accuracy are difficult to estimate, it should generally be possible to determine the major cost drivers and the leading sources of error.

There are two other modifications to the Census Bureau's list of topics that would have facilitated setting priorities. First, it would have been helpful if the list had been separated into candidates for evaluations and

---

[1] It is useful to note here that the cost of the 2010 census is projected to be over $11 billion, which is approximately $100 per housing unit. Therefore, the use of any alternatives that have substantial cost savings is a crucial benefit in looking toward the 2020 census.

candidates for formal experiments. An experiment is, generally speaking, not possible until a reasonable alternative has been identified. Therefore, the listing of any alternative methodologies along with any knowledge of their potential advantages and disadvantages will facilitate the discussion of which issues should be focused on for either experimentation or evaluation. Second, a summary of the current state of research on some of the issues described would have been helpful (in Appendix A, the column on "new to census" is related to this). While some of these issues are extremely new, some, for example questionnaire design, are topics for which the Census Bureau has a history of relevant research. This information would have supported a more refined judgment of the likelihood that use of various alternative approaches might lead to important improvements.

## 2–B PRIORITY TOPICS FOR EXPERIMENTATION IN 2010

So, without an overall strategy for the design of the 2020 census, it was difficult for the panel to develop strict priorities for the topics that should and should not be examined through the use of experiments in the 2010 census. This lack of a strategy could have been overcome to some degree with information on the potential impact on census costs and accuracy of replacing various census component processes with alternative processes. This is so because the overall goal of research on census methods has at its most basic level two main objectives: reducing costs and improving accuracy. However, this information is not available at this point and so the panel developed the following set of priority topics for experiments based on speculations concerning the possible designs of the 2020 census and qualitative information on the potential impact on costs and accuracy from the use of alternative processes. In the same vein, the primary goal of each experiment that we are recommending for priority consideration is to better understand the impacts on both census costs and census data quality resulting from the use of alternatives to current census methodology.

The three recommendations in this chapter on experimentation should be considered by the Census Bureau as the three highest priority recommendations in this report. Throughout, the panel was mindful of the special context that the decennial census provides for experimentation, and therefore one additional criterion applied was whether experimentation for the topic under consideration would substantially benefit from a decennial census environment.

To start, we put forward two topics for experimentation that were not given sufficient prominence in the list provided by the Census Bureau (see Appendix A).[2] Internet data collection was not mentioned in the list, and

---

[2] Recall that the Census Bureau typically refers to a census experiment as a study involving

the use of administrative records was mentioned very briefly (items A.6 and C.6 in Appendix A) as possibly playing a role in augmenting coverage measurement data collection, in otherwise identifying coverage problems, and in identifying and classifying duplicates. These are both very important mechanisms for improved data collection and improved evaluation.

Before expanding on those two issues, we also mention that research and experimentation on the American Community Survey (ACS) were not mentioned prominently in the 2010 Census Program for Evaluations and Experiments (CPEX) plan. We understand that ACS research and testing are intended to be handled separately, possibly using an experimental methods panel to identify improvements in ACS methodology. However, there are important commonalities between the effectiveness of methods used to collect ACS data and the methods used to collect decennial census data that need to be exploited. It is very likely that more efficient and better research will be possible by combining perspectives from both operations. An explicit recognition of both the crucial need for an ACS research and experimentation program (this is recommended in National Research Council, 2007) and the potential for cross-fertilization of such an ACS program with the CPEX program would be extremely desirable. Furthermore, given that the ACS and the decennial census will be collecting data simultaneously, measurement of the possible impact of the ACS on decennial census data collection, especially coverage follow-up (CFU) and possibly the coverage measurement effort, would be worthwhile. Finally, as we discuss below, the possible impact of the different residence concepts used by the census and the ACS is a major concern that can and should be assessed as part of the 2010 CPEX.

## 2–B.1   Internet Data Collection

The Internet is becoming the preferred mode for many households to conduct their banking, shopping, tax filing, and other official communications and interactions. It is anticipated that the Internet will also soon become a major medium for survey data collection. In the decennial census context, the Internet provides important advantages, including alternate ways of representing residence rules, increased facility for the presentation of questionnaires in foreign languages, real-time editing, and immediate transmission of data, which has important benefits for minimizing the overlap of census data collection operations. With respect to the representation of residence concepts, an Internet-based questionnaire could make it easier to display (and link to) additional examples and instructions for determining

---

field data collection—typically carried out simultaneously with the decennial census itself—in which alternatives to census processes currently in use are assessed for a subset of the population. Census evaluations are usually post hoc analyses of data collected as part of the decennial census process to determine whether individual steps in the census operated as expected.

census residence; it could also guide respondents through a more detailed set of probe questions in order to more accurately determine household counts. An Internet option could provide linguistically challenged respondents with a wider array of questionnaire assistance tools and, perhaps, administration of the actual census questions in more languages than has been feasible under the financial and logistical constraints of paper administration.

The experience in many other countries (see Appendix B for details) is that this alternative mode of response provides important benefits, which are likely to increase as 2020 advances. In particular, the recent 2006 Canadian experience is that the use of the Internet as a response option does improve the quality and timeliness of responses (Statistics Canada, 2007).

As described in Appendix B, the Census Bureau has decided against the use of the Internet in 2010 for two principal reasons. First, it believes that it is unlikely to appreciably improve the rate of response given the results of the 2003 and 2005 National Census Tests. Second, there are issues related to security that need to be considered, including the potential for hackers to disrupt the data collection, in addition to any public perception problems that are related to security concerns.[3]

It is not our charge to evaluate the Census Bureau's decision not to use the Internet for data collection in the 2010 census. However, it is obvious from the discussion in Appendix B that many countries are already strongly moving in this direction. More importantly, given the advantages listed above and the anticipation of greater advantages in the future, the Census Bureau needs to start now to prepare for use of the Internet as a major means for data collection in the 2020 census. An important step in this preparation is the inclusion of an experiment on Internet data collection in the 2010 census.

Regarding possible problems in access to and use of the Internet, the panel thinks that there may be alternative ways of interfacing with respondents that could facilitate Internet response, rather than using the mailed questionnaire as the initiating event. Regarding security concerns, Canada and other countries have been able to successfully mitigate security concerns, and it thus seems likely that the United States should be able to address this issue in time for 2020.

While the testing of an Internet response option does not require a census context, a census context would be very useful, since complex counting rules, needed for unduplicating double counts, are more easily implemented

---

[3] We note that there is generally little concern about biases in responses received by the Internet, for two reasons. First, there will always be multiple modes for response in the census given the heterogeneous population that is being counted. So mode bias is ubiquitous. Second, mode bias for the questions on the census short form will be relatively modest since there is little room for interpretation, except possibly for residence rules and race/ethnicity.

in a complete count operation. Also, response frequency is substantially higher in the census than in test censuses.

We therefore recommend that the Census Bureau include an experiment during the 2010 census that uses alternative mechanisms to facilitate Internet responses and measures the frequency of use for each, along with expeditiousness and quality of response. It may also be possible to ask the respondent if he or she would utilize an online foreign language version if available.

> *Recommendation 1:* The Census Bureau should include, in the 2010 census, a test of Internet data collection as an alternative means of enumeration. Such a test should investigate means of facilitating Internet response and should measure the impact on data quality, the expeditiousness of response, and the impact on the use of foreign language forms.

## 2–B.2   Use of Administrative Records to Assist in Component Census Operations

Administrative records are data collected as a by-product of the management of federal, state, or local governmental programs. Key examples for census applications include tax records, welfare records, building permit records, Medicare data, birth and death records, and data on immigration and emigration. Administrative records have a number of potential applications in the decennial census. These applications can be separated into those in which administrative records data are used indirectly and those in which administrative records data are used directly as decennial census data. Applications in which administrative records data are used indirectly include:

- *for improvement of the Master Address File (MAF):* addresses found in a merged administrative records file that were not on the MAF could be visited for field validation.

- *to validate edit protocols:*[4]   edit protocols that were used to make decisions about inconsistent information in responses could be based on (or evaluated using) administrative records. For example, a 22-year-old listed as living with his parents and in a prison could have his enumeration moved to the prison address through information found in administrative records.

- *for coverage improvement:* for households or individuals found on possibly more than one administrative list who were not enumerated in

---

[4]An edit protocol is an automated rule that either generates an imputed response or changes a collected response based on the values of other responses.

the census, fieldwork could be instigated at the indicated address; furthermore, addresses identified as being vacant could be checked to see if that assessment agrees with information in administrative records.

- *for coverage measurement and coverage evaluation:* consistent with A.6 in Appendix A, administrative records could be used to improve the information collected in postenumeration survey interviews;[5] furthermore, administrative records could be used to allocate demographic analysis estimates[6] to subnational regions;

- *to help target households for various purposes* (see below).

- *for duplicate search:* administrative records could be used to determine whether two records that have been matched actually represent the same person or to determine where the correct census residence is without resorting to fieldwork.[7]

Applications in which administrative records data are either used directly in the decennial census or in assessing coverage include:

- *as an alternative to last-resort proxy response:* instead of asking a neighbor or landlord for information in situations in which a respondent is not located after six attempts, if information is available from administrative lists, that information could be used for the enumeration.

- *as an alternative to item and unit imputation:* in the situations in which the Census Bureau uses either item or unit imputation (see National Research Council, 2004a, for a discussion of when unit imputation was used in the 2000 census), information from administrative records could be used as input to the imputation.

- *as a means for coverage evaluation:* whereby a person that appears on two or three administrative lists and not in the census is proof of a census omission.

In each of these applications, there could potentially be important benefits for the 2020 census, either in reducing field costs or in improving the quality of census data. We justify our optimism about the potential for applying administrative records to improve the above census component operations, and therefore the need to test those applications in the 2010 census,

---

[5]A postenumeration survey is a survey taken after the census is concluded that is used to measure coverage errors.

[6]Demographic analysis is an accounting scheme, roughly births plus immigrants minus deaths minus emigrants, for estimating the size of national demographic groups.

[7]An evaluation of A.C.E. Revision II estimates of duplication in Census 2000 using administrative records information demonstrated the potential for use of this information (for details, see Mule et al., 2007). Administrative records might be used to confirm whether enumerations that are linked by computerized search are the same persons when fieldwork was unable to provide confirmation.

given the following considerations. First, there is clearly much useful information contained in various administrative records. The nonsurvey nature of the data collection gives a real chance of being able to provide useful information on hard-to-count individuals. This advantage probably motivated the Census Bureau to attempt to use information from administrative records for coverage improvement, as in 1980 with the Non-Household Sources Check, and in 1990 with the Parolees and Probationers Check. Also, the Census Bureau will be using administrative records to generate some of the coverage follow-up interviews in 2010. However, there are also deficiencies in administrative records, including the general quality and timeliness of the information, a lack of information on race/ethnicity, a lack of current addresses, and a lack of high-quality unique identifiers. Some of the existing research has been on the use of administrative records as an alternative to taking a census, notably AREX 2000, which is not that useful in assessing the value of administrative records for census component operations. As mentioned previously, the population coverage for the more thorough of the schemes tested in AREX 2000 was between 96 and 102 percent relative to the Census 2000 counts for the five test site counties. However, the AREX 2000 and the census counted the same number of people at the housing unit level for only 51.1 percent of the households, and they counted within one person of the census for only 79.4 percent of the units.

However, the Census Bureau has made substantial progress on administrative records since then. For example, E-StARS,[8] the Census Bureau's name for a merged and unduplicated list of individuals from several administrative lists, was used to explain 85 percent of the discrepancies between the Maryland Food Stamp Registry recipients and estimates from the Census Supplementary Survey in 2001 (the pilot American Community Survey).

Although there has been much progress in collecting a higher quality merged unduplicated list of individuals, there has been little research on the nine applications listed here, in which the objective is to use administrative records not as a surrogate census but to assist in carrying out specific component operations. The panel's optimism is based not only on the information contained in administrative records, but also on the recognition that some of the component operations, especially last-resort enumeration, are understandably error-prone or are expensive (e.g., the coverage follow-up interview). Given that, administrative records do not have to be flawless to potentially provide a benefit. In addition, looking toward 2020, the quality of administrative records has been steadily improving over time. E-StARS, the Census Bureau's merged list of unique administrative records for indi-

---

[8] E-StARS is a nationwide multipurpose research database, which combines administrative records from a variety of federal and state government sources and commercial databases with micro-data modeling to produce statistics for housing units and individuals that are comparable to decennial census results.

viduals and housing units, has about the right number of people. Also, the economic directorate of the Census Bureau has been using information from administrative records directly in establishment surveys for a long time. So there is reason for optimism that some of the applications listed could be substantially improved through the use of administrative records.

It is therefore important to determine, through either experiments or evaluations, which of the above (and other) applications of administrative records are most likely to be beneficial in the 2020 census, what needs to be done to implement such techniques nationally, and what the risks and benefits are. The basic idea would be to select several counties, merge and unduplicate all the relevant lists that can be collected for both individuals and addresses in those areas, and use the information from the merged file for some of the above purposes in comparison with the current census processes. In some cases, field verification would be needed to produce metrics for comparison—which is the main reason why this might fall into the experimentation rather than the evaluation category. However, in many cases much could be discovered without additional field data collection. Clearly, a census context is extremely helpful or essential for some of the above applications, such as for duplicate search. An additional complication is that administrative records are improving in quality year by year, and therefore any experiment or evaluation should take this possibility into account. (This suggestion is closely related to items C.2 and C.6 on the Census Bureau's list of issues.)

A particular means by which administrative records could be used to reduce field costs, at the price of possibly only a negligible reduction in data quality, is targeting. Targeting is the application of a census procedure to only a subset of the population. This subset of the population is selected through use of an algorithm that attempts to differentiate between people or households that are and are not likely to benefit from the application of the procedure. This algorithm is often supported by some external data source, and, in particular, administrative records should be studied as potentially playing this role. Administrative records offer opportunities to increase the scope and effectiveness of targeting, and in particular they may have important advantages for enumerating hard-to-count populations. (In a sense, the Census Bureau already uses targeting in several respects, including targeting of the advertising campaign, targeting areas for placement of "Be Counted" forms, and targeting areas for so-called blitz enumeration techniques.)

Of course, any time one does not use a census enumeration process on some areas that is used elsewhere, some of the omitted areas may have slightly poorer quality data as a result. So, for example, if a block canvass is not used in a particular block, there is a chance that new housing units there will be missed and that the area will receive a lower count as a result. (It should be noted that the Census Bureau has previously considered targeting for use with block canvassing, but to this point it has rejected this idea.) However, if properly planned and implemented, targeting should increase overall census data accuracy and at the same time reduce costs. This is because, if the targeting is effective, the reduction in data quality due to the selective omission of a census process is likely to be very slight. The resources saved through the use of targeting can then be used in other ways to improve the overall census data quality. Furthermore, sometimes resources are already constrained, and for those situations the question may not be whether to use targeting, but how best to use it. Also, through use of an algorithm, there is no intentional bias against any given area. (It may also be worth mentioning that some suggest that targeting can be perceived as uncomfortably close to sampling for the count. This is clearly an incorrect perception; it is merely the allocation of scarce resources to those cases most likely to benefit from this additional effort at enumeration.)

Clearly, further research (either experimentation or evaluation) is needed before targeting can be used in the decennial census. Given the promise of targeting, the panel thinks that the Census Bureau should prioritize either experimentation or evaluations that assess the promise of various forms of targeting and therefore retain sufficient data to ensure that such evaluations can be carried out. (Targeting is included in items C.3 and E.2 on the Census Bureau's list.) Creation of a Master Trace Sample, discussed in Chapter 3, is likely to satisfy this data need.

> *Recommendation 2:* The Census Bureau should develop an experiment (or evaluation) that assesses the utility of administrative records for assistance in specific census component processes—for example, for improvement of the Master Address File, for nonresponse follow-up, for assessment of duplicate status, and for coverage improvement. In addition, either as an experiment or through evaluations, the Census Bureau should collect sufficient data to support assessment of the degree to which targeting various census processes, using administrative records, could reduce census costs or improve census quality.

### 2–B.3  Alternative Questionnaire Experiment

The 1980, 1990, and 2000 censuses have all involved some type of alternative questionnaire experiment in the associated research programs. The reason is straightforward: anything that can be done to increase response to questionnaires when they are sent out will necessarily decrease the amount of work that must be done by enumerators in the field in following up with nonrespondents. Also, to the extent that the initial questionnaire can be made clear, the quality of the collected data should improve. It is therefore of high priority that an alternative questionnaire experiment should be employed in the 2010 CPEX.

The Panel on Residence Rules in the Decennial Census (National Research Council, 2006:Finding 8.2) observed that "the Census Bureau often relies on small numbers (20 or less) of cognitive interviews or very large field tests (tens or hundreds of thousands of households, in omnibus census operational tests) to reach conclusions about the effectiveness of changes in census enumeration procedures." That panel argued for the development of more mid-range, smaller scale tests. We concur; there are numerous questionnaire design issues for which smaller scale tests would be a preferable vehicle compared with a formal census experiment. In thinking about an alternative questionnaire experiment or experiments for the 2010 census, the question is: Which sets of possible changes to the census questionnaire most need (or would most benefit) from being conducted in the census environment?

### 2–B.4   Race/Ethnicity as a Single Question

On page 1 of the short-form-only questionnaire planned for use in the 2008 census dress rehearsal (see Figure 2-1), the two questions on race and Hispanic origin (questions 8 and 9) take up half of the second column and about 40 percent of the respondent-fillable space on the page. Likewise, the race and Hispanic origin questions take up about half of the space allotted to collect information on persons 2 through 6 in a household (the block for Person 2 is shown in Figure 2-2). In the short-form-only census planned for 2010, then, the largest share of the questionnaire is given to the questions on race and Hispanic origin; therefore, if a viable alternative exists, a major focus of a questionnaire experiment in the 2010 census should be one that focuses on the two questions on race and ethnicity, since the rate of response is typically associated with the perceived ease of compliance.

Information on race is currently requested on the census questionnaire in response to the needs of the Voting Rights Act of 1965. In 1997, the Office of Management and Budget (OMB) developed standards for racial and ethnic classification to be used in the 2000 census, which resulted in 63 possible responses to account for multiple race identification. These standards will continue to apply to the 2010 census. Ethnicity, defined as either "of Hispanic origin" or "not of Hispanic origin," was requested on a separate question in the 2000 census, resulting in 126 total race/ethnicity response categories.

Evaluations have shown that the race/ethnicity questions used in 2000 (and in previous censuses) were associated with substantial confusion of race and ethnicity, often resulting in nonresponse, in some (seemingly) contradictory responses to the decennial census questions, and in high frequencies of

United States Census 2010

**2008 Dress Rehearsal**

This is the official form for all the people at this address.
It is quick and easy, and your answers are protected by law.

U.S. DEPARTMENT OF COMMERCE
Economics and Statistics Administration
U.S. CENSUS BUREAU

**Draft**

**Use a blue or black pen.**

***Start here***

**The Census must count every person living here on April 1, 2008.**

**Before you answer Question 1, count the people living in this house, apartment, or mobile home using our guidelines.**

- Count all people, including babies, who live and sleep here most of the time.

**The Census Bureau also conducts counts in institutions and other places, so:**

- Do not count anyone living away either at college or in the Armed Forces.
- Do not count anyone in a nursing home, jail, prison, detention facility, etc., on April 1, 2008.
- Leave these people off your form, even if they will return to live here after they leave college, the nursing home, the military, jail, etc. Otherwise, they may be counted twice.

**The Census must also include people without a permanent place to stay, so:**

- If someone who has no permanent place to stay is staying here on April 1, 2008, count that person. Otherwise, he or she may be missed in the census.

**1. How many people were living or staying in this house, apartment, or mobile home on April 1, 2008?**

***Number of people =***

**2. Were there any additional people staying here April 1, 2008 that you did not include in Question 1?**
*Mark ✗ all that apply.*

Children, such as newborn babies or foster children
Relatives, such as adult children, cousins, or in-laws
Nonrelatives, such as roommates or live-in baby sitters
People staying here temporarily
No additional people

**3. Is this house, apartment, or mobile home —**
*Mark ✗ ONE box.*

Owned by you or someone in this household with a mortgage or loan? *Include home equity loans.*
Owned by you or someone in this household free and clear (without a mortgage or loan)?
Rented?
Occupied without payment of rent?

**4. What is your telephone number?** *We may call if we don't understand an answer.*
Area Code + Number

**5. Please provide information for each person living here. Start with a person living here who owns or rents this house, apartment, or mobile home. If the owner or renter lives somewhere else, start with any adult living here. This will be Person 1.**
**What is Person 1's name?** *Print name below.*

Last Name

First Name MI

**6. What is Person 1's sex?** *Mark ✗ ONE box.*

Male Female

**7. What is Person 1's age and what is Person 1's date of birth?**
*Please report babies as age 0 when the child is less than 1 year old.*
*Print numbers in boxes.*

Age on April 1, 2008 Month Day Year of birth

→ **NOTE: Please answer BOTH Question 8 about Hispanic origin and Question 9 about race. For this census, Hispanic origins are not races.**

**8. Is Person 1 of Hispanic, Latino, or Spanish origin?**

No, not of Hispanic, Latino, or Spanish origin
Yes, Mexican, Mexican Am., Chicano
Yes, Puerto Rican
Yes, Cuban
Yes, another Hispanic, Latino, or Spanish origin — *Print origin, for example, Argentinean, Colombian, Dominican, Nicaraguan, Salvadoran, Spaniard, and so on.*

**9. What is Person 1's race?** *Mark ✗ one or more boxes.*

White
Black, African Am., or Negro
American Indian or Alaska Native — *Print name of enrolled or principal tribe.*

Asian Indian
Chinese
Filipino
Japanese
Korean
Vietnamese
Native Hawaiian
Guamanian or Chamorro
Samoan
Other Asian — *Print race, for example, Hmong, Laotian, Thai, Pakistani, Cambodian, and so on.*
Other Pacific Islander — *Print race, for example, Fijian, Tongan, and so on.*

Some other race — *Print race.*

**10. Does Person 1 sometimes live or stay somewhere else?**

No Yes — *Mark ✗ all that apply.*

In college housing
In the military
At a seasonal or second residence
For child custody
In jail or prison
In a nursing home
For another reason

→ **If more people were counted in Question 1, continue with Person 2.**

OMB No. 0607-0915: Approval Expires x/xx/xxxx

Form DX-1 (4-24-2007)

USCENSUSBUREAU

**Figure 2-1** First page (Person 1), draft 2008 dress rehearsal questionnaire

SOURCE: http://www.census.gov/Press-Release/www/2007/questionnaire_4_24_07.pdf.

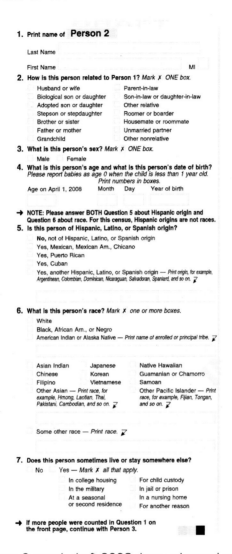

**Figure 2-2** Person 2 panel, draft 2008 dress rehearsal questionnaire

SOURCE: http://www.census.gov/Press-Release/www/2007/questionnaire_4_24_07.pdf.

response of "some other race" for Hispanic respondents (see, e.g., Census 2000 Topic Report #9, Race and Ethnicity in Census 2000, Census 2000 Testing, Experimentation, and Evaluation Program). Over the past 20 years, the Census Bureau has devoted considerable research to testing various approaches to the design of questions on race and ethnicity, trying alternative question wordings, formatting, and sequencing to elicit quality information (see, e.g., Rodriguez, 1994; McKay and de la Puente, 1995; de la Puente and McKay, 1995),

The Census Bureau has included race/ethnicity as one of their 11 topic groups for possible experimentation or evaluation in 2010. However, the Bureau gives low priority to the issue of developing a combined race and ethnicity question (listed as item B.2 in Appendix A). We disagree with that assessment; race and ethnicity are not really separate notions for many respondents, and the confusion resulting from the use of separate questions might be substantially reduced through the use of a single race/ethnicity question. This notion has been previously tested by the U.S. Census Bureau (1997) with generally positive results. Furthermore, the tendency to report "some other race" rather than Hispanic is likely to be reduced through the use of a single question.

The current race and ethnicity questions provide a number of examples of specific groups, including Filipino, Guamanian, or Samoan for race, and Puerto Rican and Cuban for ethnicity. There is no legal obligation stemming from the Voting Rights Act for the census questionnaire to include the mention of these various specific groups on the census short form. The argument in favor of including as many groups as the form will support is that this may increase response given personal feelings of affiliation with very specific groups. Also, some argue that use of a streamlined questionnaire—that is, one that does not mention these individual groups—will increase the frequency of the mistaken response of "some other race." However, we suspect that the response of "some other race" is much more a function of the separation of race and ethnicity into two questions. Furthermore, we think that the inclusion of the specific groups makes the entire census questionnaire appear more complex, which may lower the response rate. We acknowledge that there is great interest in the relative size of these numerically smaller race and ethnic groups for states and counties, but that information will now be available on the American Community Survey.

We therefore think that the Census Bureau should include, as an experiment, the use of a single question on race and ethnicity. In addition, a streamlined version of this should also be tested, in which the only groups listed are (1) white, (2) black, (3) American Indian or Alaskan Native, (4) Asian, (5) Native Hawaiian or Pacific Islander, and (6) Hispanic, allowing for multiple responses in all of these categories.[9]    We think that this is a productive avenue for testing because of its potential improvement regarding data quality. However, progress will be difficult, since the best approach to collecting higher quality data without discouraging respondents is not obvious. Continued experimentation is therefore imperative.

Finally, in addition to the test of a single race/ethnicity question, in-depth follow-up of a small sample of individuals who provide inconsistent responses to the 2010 questions should be planned.[10]    Without understanding respondent behavior induced by a given question wording, it is very difficult to come up with hypotheses about how to improve that wording. Therefore, it would be useful to contact 50 or 100 such individuals and through face-to-face interviews determine why they responded the way that they did.

### 2–B.5    Representation of Residence Concepts

In terms of physical space on the page, the items on race and ethnicity take up the greatest area due to the number of responses permitted. However, the largest single presentation of a question has been Question 1 on recent censuses: the count of residents at the household.

The 2010 census will follow the basic concept laid out in the law authorizing the first census in 1790 of counting people at their "usual place of abode" (1 Stat. 105). Over time, this concept has evolved into one of counting people at their usual residence; this is distinct from counting them at their current residence or the location where they are when reached by the census. The Census Bureau has developed sets of residence rules to determine how to handle cases in which residential location may be ambiguous. Since the switch to reliance on the mail for most census data collection, the phrasing of Question 1 and the instructions that accompany it have been continually revised in order to guide census respondents to reporting their own residential situation in a way that is consistent with the Census Bureau's residence rules.

The National Research Council (2006) report *Once, Only Once, and in the Right Place: Residence Rules in the Decennial Census* comprehensively reviewed census residence rules past and present, assessing their adequacy

---

[9]It should be noted that this specific question format runs counter to a provision included in the fiscal year 2005 omnibus appropriations bill (and that was made binding on subsequent years), which requires the Bureau to include a "some other race" option.

[10]Inconsistency is by necessity apparent since the responses for children with parents of different races or ethnicities may not be clear and, more importantly, since race and ethnicity responses are a matter of self-identification that does not need to be consistent. Apparently inconsistent responses include respondents who check a category indicating that they consider themselves to belong to a specific Hispanic group but at the same time also responding that they are not of Hispanic, Latino, or Spanish origin.

in light of societal changes that can complicate clear definition of residence. These changes include the growth of both "sunbird" and "snowbird" populations that move to different areas based on seasonal weather changes, the changing nature of family structures (including children in shared custody arrangements), and the emergence of assisted living facilities for the elderly. The 2006 report also considered long-standing historical challenges to accurate residence measurement, particularly concerning the large share of the nonhousehold (or group quarters) population living in places like college dormitories and correctional facilities.

Based on its review, the study panel suggested additional areas of research. Primary among these was a call to collect "any residence elsewhere" information: allowing respondents to specify a specific street address for another location at which they consider themselves a resident, as well as a follow-up question about whether the respondent considers this other location to be their "usual residence" (National Research Council, 2006:Rec. 6.2). That panel specifically suggested that "any residence elsewhere" be asked of the general household population in a 2010 census experiment and that the resulting data be comprehensively reviewed in an evaluation report (National Research Council, 2006:Recs. 6.5, 8.4). It also suggested that the "any residence elsewhere" question be asked of all group quarters respondents in 2010 (National Research Council, 2006:Sec. 7–D); a similar "usual home elsewhere" question was asked on all group quarters questionnaires in 2000, but they were processed and considered valid only for particular group quarters types.

A major reason for the importance of collection of "any residence elsewhere" information on a test basis for the general population is to help resolve a major outstanding concern about the transition from the traditional census long form to the ongoing American Community Survey. While the decennial census uses a "usual residence" concept, the ACS uses something closer to a "current residence" rule; specifically, residence in the ACS is defined using a "two-month rule" relative to the time of interview (see National Research Council, 2006:Box 8-2 and Sec. 8–C for extended discussion). The differences in census and ACS estimates that may be attributed to their differing residence standards is as yet unknown and is a concern on which solid data are critically important. To that end, National Research Council (2006:Rec. 8.3) suggested the twofold approach of testing the "any residence elsewhere" question in the 2010 census and testing a "usual residence"-type question on the ACS questionnaire as a separate ACS research activity.

In addition to the "any residence elsewhere" query, National Research Council (2006:Rec. 6.5) suggested that additional methods for presenting residence rules and concepts be included in a 2010 alternative questionnaire experiment. In particular, the panel suggested a shift away from the model of lengthy instructions before Question 1 and instead breaking the resident question into smaller, easier-to-parse questions. This work could build on alternative questionnaire presentations that the Census Bureau tested on a limited basis in its 2005 National Census Test and an ad hoc test in 2006. To be clear—and as is noted elsewhere in this report—National Research Council (2006) argued that the Census Bureau often relies too much on both very small and very large tests, and that some residence-related questions (e.g., specific cues to include on questionnaires or alternative means of developing rosters of household members) may be better handled by other testing means. However, the importance of Question 1, the potential gain in data accuracy, and the potential reduction in the need to dispatch an enumerator to conduct a coverage follow-up interview that could stem from even small changes on the question form all argue strongly for a residence component of a 2010 alternative questionnaire experiment.

## 2–B.6    Other Content Issues

Other content issues on the 2010 census form are also worth examining and might benefit from an experiment in 2010. The hope is that these various questionnaire wording issues could be folded in with an experiment on race and ethnicity, residence rules, or both. There may be too many issues for a single experiment and therefore there may be a need to further prioritize these issues before finalizing an alternative questionnaire experiment.

### Coverage Probes

Two coverage probes will be included on the 2010 census questionnaire for the first time. These are: (1) "Were there any additional people staying here April 1, 2010 that you did not include in Question 1?" and (2) "Does Person X sometimes live or stay somewhere else?" This is followed by a listing of situations that are sometimes reported in error. As implemented in 2010, this set of probes is primarily intended as a trigger for inclusion in the coverage follow-up operation, described below. The probes also serve to jog a respondent's memory and prompt them to reevaluate their answer to the household resident count in Question 1 of the census form. It is worth considering whether more specific or differently worded probes are more effective at accomplishing either of these tasks, and whether they can be structured to provide auxiliary information that could be useful in editing census responses. For instance, a more detailed query about whether the respondent is at (or may be counted at) a seasonal residence, or a focused question on the residence of college-enrolled children, may prove to have advantages over the approach planned for 2010.

## Motivation of Respondents

The 2006 Canadian census questionnaire added brief descriptive statements at key places in order to anticipate respondents' concern about a question's justification in the census. By including these, Statistics Canada thinks that it has achieved some benefits in building respondent motivation to answer questions on the census form. For example, the 2006 census long-form questions on race and ancestry—which, in Canada, are not part of the short-form questions asked of everybody—are prefaced with the explanation:

> The census has collected information on the ancestral origins of the population for over 100 years to capture the composition of Canada's diverse population.

The specific race question includes the reminder that this information is collected to support programs that promote equal opportunity for everyone to share in the social, cultural, and economic life of Canada.

The last page of the Canadian short-form questionnaire includes a paragraph-length section labeled "Reasons Why We Ask the Questions," noting, for example, that "Question 7 on languages is asked to implement programs that protect the rights of Canadians under the Canadian Charter of Rights and Freedoms. It also helps determine the need for language training and services in English and French." It could be useful to measure the impact on the quality of response that would result from various attempts to represent similar motivational messages on the U.S. census form.

## Group Quarters

Given that some types of group quarters' residences are subject to a high rate of duplication, in particular those in college dormitories (see Mule, 2002), it might be useful to evaluate the benefits of a "usual home elsewhere" question on the census questionnaire for all types of group quarters residences. (This is consistent with Recommendation 6.2 and Section 7-C in National Research Council, 2006.) This might facilitate real-time identification of census duplicates between residents of group quarters and residents of nongroup quarters.

Finally, item G.1 on the Census Bureau's list of research topics proposes administering the 2000 census questionnaire to a group of 2010 census respondents so that some insight can be drawn about the effectiveness of the complete bundle of changes between the 2010 and 2000 forms. This proposal to use the prior census questionnaire as a control group treatment has not always been carried out in past alternative questionnaire experiments. Implementing it is consistent with guidance from the previous National Research Council (2006:Rec. 6.8) report, and we concur that it should be done as part of a 2010 alternative questionnaire experiment.

## 2–B.7    Deadline Messaging and Other Presentation Issues

Deadline messaging includes a variety of ways of notifying the respondent on mailing materials that in order to be accepted the enclosed questionnaire has to be returned by a given date. By a compressed mailing schedule is meant that, instead of the approach used in the 2000 census, in which the questionnaire was mailed two weeks before Census Day, the households will receive the census questionnaire just a few days before Census Day. In the 2006 decennial short-form experiment,[11] the use of deadline messaging, in conjunction with a compressed mailing schedule, resulted in a higher mail response rate (Martin, 2007). The deadline message was placed on the advance letter informing the household of the upcoming appearance of the census questionnaire, on the envelope of the initial mailed questionnaire, on the initial questionnaire cover letter, and on the reminder postcard. However, the 2006 test could not determine whether the increased response was due to a specific form of the deadline message or whether it was due to the compressed mailing schedule. Therefore, some further work attempting to determine the specific cause of the increase in response would be extremely useful. More importantly, since increasing the initial response rates decreases the nonresponse follow-up fieldwork, which reduces census costs, this is important to investigate further. Additional research on the effectiveness of different dates for both the initial mailing of the census questionnaires and the mailing of the replacement questionnaires would also be useful to undertake. Item H.1 on the Census Bureau's list argues that looking at this issue in a census environment is important, and the panel agrees, since response to mail materials differs in a census in comparison to either a test census or a survey environment.

We have described a number of issues that relate to the content and the presentation of the census questionnaire, including race and ethnicity, residence rules, coverage probes, providing a motivation for the cooperation of respondents, collection of alternate address data for residents of group quarters, and deadline messaging. It may be that several of these issues can be jointly addressed in a single experiment by including these issues as separate factors in the experiment. One straightforward way of accomplishing this, which is much more cost-effective with respect to the burden on respondents, is through the use of a fractional factorial design, assuming that some of the higher level interactions between these factors are negligible (see Box and Hunter, 1961).

---

[11]The decennial short-form experiment evaluated several potential improvements to the census mail form. These included a revised instruction about whom to list as Person 1, a series of questions to reduce and identify coverage errors, and a deadline for return of the form.

*Recommendation 3:* The Census Bureau should include one or more alternate questionnaire experiments during the 2010 census to examine:

- the representation of questions on race and ethnicity on the census questionnaire, particularly asking about race and Hispanic origin as a single question;
- the representation of residence rules and concepts on the census questionnaire; and
- the usefulness of including new or improved questions or other information on the questionnaire with regard to (1) coverage probes, (2) the motivation of census questions, (3) the request of information on usual home elsewhere on group quarters questionnaires, and (4) deadline messaging and mailing dates for questionnaires.

In such experiments, both the 2000 and the 2010 census questionnaires should be included in the assessments. The Census Bureau should explore the possibility of joining the recommended experiments listed above into a single experiment, through use of fractional factorial experimental designs.

### 2–B.8  A Possible Additional Experiment: Comparison of Telephone to Personal Interview for Coverage Follow-Up Interview

The current plans are to carry out a coverage follow-up interview in 2010 to collect additional information for six situations for which the number of residents is unclear based on the responses to the initial questionnaire (see Box 2-1). Since a large fraction (probably more than 20 percent) of U.S. households may satisfy one or more of these six situations, the costs of the resulting coverage follow-up interviews could be prohibitive. To reduce these costs, the Census Bureau is planning to follow up these households by telephone only (and therefore only for those households that provide a contact telephone number on the census questionnaire).

This specific implementation of the coverage follow-up interview raises some concerns about the quality of the information received. First, we are concerned that the households that would most benefit from this follow-up will be those not likely to provide valid telephone numbers and consequently will be missed. For example, some of those that are harder to enumerate may make use of prepaid cell phones. Therefore, it would be useful to determine whether other wordings of the request for phone numbers would increase the response to this item. (This relates to the earlier issue of providing motivation for questions on the short form. This suggestion is related to items C.8, C.7, F.1, and F.2 on the Census Bureau's list of issues.)

---

**Box 2-1**   Situations Generating a Coverage Follow-Up Interview

1. Households with discrepancies between the household counts and the number of individuals for which information is provided

2. Households with more than six residents (which will therefore not fit on the census questionnaire

3. Households that indicate on the census questionnaire other households in which the residents might also have been enumerated

4. Households that indicate other people not included in response that sometimes live there

5. Households that are identified as having individuals that might have been duplicated in the census through use of a national computer search for duplicates

6. Households that may have not been correctly enumerated given information from administrative records.

SOURCE: Adapted from information from U.S. Census Bureau; see also National Research Council (2006:Box 6-3).

---

Another concern stems from the fact that the coverage follow-up interview uses question wording similar to that on the census questionnaire, and there is thus a good chance of generating the same response as was initially received in the case of interviews resulting from coverage probes or from the identification of potential duplicates. One alternative to address this concern that might be worth examining is whether there is a way of communicating to the respondent the circumstances that generated the interview through a series of probes. A second way of addressing this concern is that higher quality answers, possibly using such probes, might be produced through use of a face-to-face interview, rather than a phone interview. While this would clearly be more expensive, knowing the impact on quality would be useful in designing the analogous data collection in 2020. Also, there are ways of reducing field interview costs to permit more face-to-face interviewing. For example, the targeting of households through the use of administrative records might reduce the workload to a manageable level, allowing for face-to-face interviews of selected households.

If the decision is made not to include study of the coverage follow-up interview in a census experiment, the above concerns strongly argue for retention of all relevant information to be able to evaluate this process after the census is completed.

## 2–C   CONCLUSION

These are the panel's suggestions for experiments to be carried out during the 2010 census. We look forward to assisting the Census Bureau in fleshing out more specific study plans for the ideas that are ultimately selected for experimentation in the coming months.

We also think that the Census Bureau needs to increase its in-house expertise in experimental design regarding census experimentation. The panel has seen evidence in the past that some experiments, in both censuses and test censuses, have not been fully consistent with accepted principles of experimental design. This includes the use of preliminary assessments of which factors might affect a response of interest, the use of controls and blocking for meaningful comparisons (see, e.g., National Research Council, 2006:Rec. 6.8), and the simultaneous varying of test factors (including use of orthogonal designs, factorial designs, and fractional factorial designs) for greater effectiveness of test panels. Also, often not enough attention is paid in advance to the statistical power of tests. Certainly some of this can be attributed to the fact that the primary function of a census or a census test is an opportunity to assess the full census operation with the embedded experiments having to make do with various limitations. However, it is important for the Census Bureau to improve its application of experimental design techniques for its experiments, both to reduce the costs of the experimentation and to increase the information contained in the results.

# – 3 –

# Initial Views on 2010 Census Evaluations

## 3–A  SUGGESTIONS FOR THE 2010 CENSUS EVALUATIONS

The panel's first priority is to provide input to the selection of experiments to be implemented in 2010, since the design of these experiments needs to begin very soon to allow for the development of associated materials and protocols. In addition, the panel has some suggestions relative to the evaluations to be carried out in conjunction with the 2010 census. There is also a time pressure for them since, as stated previously, much of the data collection in support of the 2010 census evaluations needs to be specified relatively early, in particular so that the contractors involved in many of the census processes can make plans for the collection and structuring of data extracts that relate to the functioning of those processes.

### 3–A.1  Address List Improvement

For the 2000 census, the Census Bureau departed from past practice of building the address list for the census from scratch. Instead, it pursued a strategy of building a Master Address File (MAF), using the 1990 address list as a base and seeking ways to "refresh" the database during the intercensal period. Legislation enacted in 1994 created two major tools for address list improvement. First, the new law authorized the Census Bureau to use the U.S. Postal Service's Delivery Sequence File (DSF; as the name suggests, a master list of mail delivery addresses and locations used to plan postal routes) as an input source. Second, it permitted limited sharing of extracts of the Master Address File (which is confidential information under Title 13 of the U.S. Code) with local and tribal governments. Specifically, this provision led to the creation of the Local Update of Census Addresses (LUCA) program, first conducted in several phases in 1998 and 1999 (see National Research Council, 2004a:62–65).

The Master Address File used to support the American Community Survey during the intercensal period is essentially an update of the 2000 census MAF, revised to include edits to the Postal Service's Delivery Sequence File and new construction. Through these actions, the MAF, heading into the 2010 census, will be certainly more than 90 percent complete but probably not 99 percent complete. (There will almost certainly be a substantial amount of duplication as well.)

The Census Bureau will utilize two operations to increase the degree of completeness of the MAF from its status in 2008 in preparation for its use in the decennial census in 2010. First, it will again use the LUCA program, in which local governments will be asked to review preliminary versions of the MAF for completeness and to provide addresses that may have been missed (or added in error). However, even granting that LUCA will be improved over the 2000 version, it is likely that the participation will be uneven and that a substantial amount of incompleteness will remain after these addresses are added to the MAF. In anticipation of that, the Census Bureau will carry out a national block canvass, visiting each census block, and adding any missed housing units to the MAF (while collecting information from global positioning systems for all housing units).

It may be the case that for many well-established blocks in the United States a 100 percent block canvass is wasteful, given that there is little possibility in these blocks of addition or deletion of housing units over time. It would be useful to identify such blocks in advance, since then the block canvass could be restricted to the subset of blocks in need of MAF updating (this is consistent with item C.3 in Appendix A). Given the costs of a 100 percent block canvass, identifying a targeting methodology that does an excellent job of discriminating between those blocks that are very stable over time and those blocks that are likely to have recent additions or deletions (or both) would provide substantial cost savings with possibly only a negligible increase in the number of omissions (or erroneous inclusions) in the MAF. It is likely that administrative records, especially building permit records, commercial geographic information systems, and the ACS could provide useful predictors in discriminating between stable and nonstable blocks. Such targeting is already used in the Canadian census; it uses an address register that is updated intercensally, and field verification is restricted to areas where building permit data indicate the presence of significant new construction (Swain et al., 1992).

To support the determination as to whether any targeting methods might satisfy this need—and, indeed, to facilitate a richer evaluation of MAF accuracy than was possible in 2000—the Census Bureau should ensure that the complete source code history of every MAF address is recoverable. In 2000, the MAF was not structured so that it was possible to fully track the procedural history of addresses—that is, which operations added, deleted, or modified the address at different points of time. Therefore, it was not possible to accurately determine the unique contributions of an operation like LUCA or the block canvass; nor was it possible to assess the degree to which various operations overlapped each other in listing the same addresses. Census Bureau staff ultimately derived an approximate "original source code" for MAF addresses, albeit with great difficulty; see National Research Council (2004b:146–147). Redesign of the MAF database structure was included in the plans to enhance MAF and TIGER during this decade; the Census Bureau should assess whether the new structure will adequately track the steps in construction of the 2010 (and future) MAF.

> *Recommendation 4:* **The Census Bureau should design its Master Address File so that the complete operational history—when list-building operations have added, deleted, modified, or simply replicated a particular address record—can be reconstructed. This information will support a comprehensive evaluation of the Local Update of Census Addresses and address canvassing. In addition, sufficient information should be retained, including relevant information from administrative records and the American Community Survey, to support evaluations of methods for targeting blocks that may not benefit from block canvassing. Finally, efforts should be made to obtain addresses from commercial mailing lists to determine whether they also might be able to reduce the need for block canvassing.**

### 3–A.2 Master Trace Sample

The idea of creating a master trace sample, namely designating a sample of households in, say, census blocks, for which the full history of relevant census operations is retained in an accessible manner for subsequent analysis, is extremely important. In each decennial census, there are unanticipated problems that need to be fully understood in order to make modifications to the census design, to partially or completely eliminate their chance of occurring in the subsequent decennial census. A master trace sample provides an omnibus tool for investigating the source of any of a large variety of potential deficiencies that can arise in such a complicated undertaking as the decennial census. Otherwise, the Census Bureau is usually left with evaluation studies that, due to the limited information available, are often univariate or bivariate summaries that cannot inform about even relatively simple interactions between the individuals, the housing unit, and the enumeration techniques that resulted in a higher frequency of coverage (or content) errors.

The value of a master trace sample database or system has been advocated by several National Research Council panels, including the Panel on Decennial Census Methodology (National Research Council, 1985:Rec. 6.3), the second phase of the Panel on Decennial Census Methodology (National Research Council, 1988), the Panel on Alternative Census Methodologies (National Research Council, 1999:Rec. 5.1), and the Panel on Research on Future Census Methods (National Research Council, 2004b:Rec. 8.4, 8.5, 8.6, 8.7). The last cited report contains a useful history of the development of this idea and includes the following recommendation: "The Census Bureau should carry out its future development in this area of tracing all aspects of census operations with the ultimate aim of creating a Master Trace *System*, developing a capacity for real-time evaluation by linking census operational databases as currently done by the Master Trace Sample. Emerging 21st century technology should make it feasible to know almost instantaneously the status of various census activities and how they interact. Such a system should be seriously pursued by the Census Bureau, whether or not it can be attained by 2010 (or even by 2020)." Such a proposal is a straightforward generalization of item A.3 of the Census Bureau's list, though expanding from a focus on the coverage measurement survey to the full set of census operations.

Such a database could be used to evaluate many things, including determining what percentage of census omissions are in partially enumerated households and what percentage of omissions are found on the merged administrative records database. A master trace sample database would be extremely useful in addressing the needs described in the previous section, including understanding the source of duplicates in the Master Address File and evaluating the benefits of LUCA and the block canvass operation. An overall assessment of the workings of the coverage follow-up interview would be feasible if the master trace sample database collected sufficient data so that it was known for each housing unit in the CFU interview what triggered the CFU interview and what the result of the interview was—that is, what changes were made and what information precipitated the change. As indicated, inclusion of the merged administrative records file and relevant data from the American Community Survey in such a database would provide additional information at the individual and local area levels.

Creation of a master trace sample presents a number of challenges. First, there is the retention of the data from the census and affiliated activities. Some modest planning is needed here, especially given the necessity of collecting data from various contractors who are likely not to have planned in advance to provide for such data extracts. In addition, it is necessary to find an effective way of linking the information retained about the enumerators, the housing units, the residents, the census processes, the type of census coverage error made, and contextual information in a way that facilitates a broad range of potential analyses, especially those that examine interactions among these various aspects of the census process. Also, selecting the minimum data to be collected that is included in the master trace sample database is crucial to address early on. This is because while the addition of various sets of variables from different parts of the census and the census management information system provides broader capabilities for investigating various aspects of census-taking, the inclusion of each additional set of variables complicates the formation of the database. This is a hard database management problem, and the Census Bureau should enter into such a project with the recognition of the need for input of considerable expertise in database management to ensure success. (We think that the relative lack of use of the 2000 Master Trace Sample was due in part to its inability to facilitate many types of analysis.)

An additional concern is that the sampled blocks included have to be kept confidential so that the behavior in these blocks is representative of the entire census. Finally, we do not think the size of the master trace sample database is a major concern. A smaller but somewhat analogous database was constructed by the Census Bureau in 2000 and, as noted above, there have been substantial advances in computing memory and speed since then.

> *Recommendation 5:* **The Census Bureau should initiate efforts now for planning the general design of a master trace sample database and should plan for retention of the necessary information to support its creation.**

### 3–A.3   Reverse Record Check

The Canadian Census has successfully employed a reverse record check for the last eight censuses to measure net coverage error. Briefly, four samples are collected: (1) a sample of enumerations from the previous census, (2) a sample of births in the intercensal period, (3) a sample of immigrants in the intercensal period, and (4) a sample of those missed in the previous census. The fourth sample is clearly the most difficult, but by matching those contained in the four samples for the previous reverse record check to the census to determine omissions and continuing this process over several censuses, a relatively useful sample of omissions can be formed over time. Once the four samples are formed, current addresses are determined, and the sample is matched to the census using name, addresses, and other characteristics. In a separate operation, the census is matched against itself to generate an estimate of the overcount, and, using both, an estimate of the net undercount is derived. Characteristics for both the omissions and overcounts support tabulations by age, sex, race, geography, etc.

To date, this procedure has not been used to evaluate the U.S. decennial census, mainly due to the 10-year period between censuses (as opposed to the 5 years between Canadian censuses), which complicates the need to trace people's addresses from one census to the next. This issue was specifically examined in the Forward Trace Study (Mulry-Liggan, 1986). However, with administrative records systems improving each year, and given the emergence of the American Community Survey, tracing people over a 10-year period is likely to be much more feasible now in comparison to 1984. Furthermore, a reverse record check has an important advantage over the use of a postenumeration survey with dual-systems estimation in that there is no need to rely on assumptions of independence or homogeneity to avoid correlation bias, a type of bias that occurs in estimating those missed by both the census and the postenumeration survey. There are also more opportunities for validating the reliability of the estimates provided. For example, a reverse record check provides an estimate of the death rate. The key issue concerning feasibility remains tracing, and a useful test of this would be to take the 2006-2007 ACS and match that forward to see how many addresses could be found over the 3.5-year period. In such a test, the ACS would serve as a surrogate for the sample from the previous census enumerations. Either relating this back to a sample of census enumerations and a sample of census omissions, or developing a sample of ACS omissions, remains to be worked out. But certainly, successful tracing of nearly 100 percent of the ACS would be an encouraging first step.

> *Recommendation 6:* **The Census Bureau, through the use of an experiment in the 2010 census (or an evaluation of the 2010 census) should determine the extent to which the American Community Survey could be used as a means for evaluating the coverage of the decennial census through use of a reverse record check.**

### 3–A.4    Edit Protocols

Edit protocols are decisions about enumerations or the associated characteristics for a housing unit that are made based on information already collected, hence avoiding additional fieldwork. For example, an edit protocol might be that, when an individual between ages 18 and 21 is enumerated both away at college and at their parent's home, the enumeration at the parent's home is deleted. (Note that census residence rules are to enumerate college students where they are living the majority of the time, which is typically at the college residence.) This would avoid sending enumerators either to the parent's home or to the college residence, but it would occasionally make this decision in error. The Census Bureau has made widespread use of edit protocols in the past to deal with inconsistent data. For example, there are rules to deal with inconsistent ages and dates of birth. Furthermore, early in 2000, when it became apparent that the MAF had a large number of duplicate addresses, the Census Bureau developed an edit protocol to identify the final count for households with more than one submitted questionnaire (see Nash, 2000).

More generally, edit protocols might be useful in resolving duplicate residences, as well as in situations in which the household count does not equal the number of people who are listed as residents. Again, as with targeting, edit protocols avoid field costs but do have the potential of increased census error. However, given the increasing costs of the decennial census, understanding precisely what the trade-offs are for various potential edit protocols would give the Census Bureau a better idea of which of these ideas are more or less promising to use in the 2020 census. The panel therefore suggests that the Census Bureau prioritize evaluations that assess the promise of various forms of edit protocols and therefore retain sufficient data to ensure that such evaluations can be carried out. Creation of a master trace sample is likely to satisfy this data need.

### 3–A.5 Coverage Assessment of Group Quarters

The census coverage measurement program in 2010 will not assess some aspects of the coverage error for individuals living in group quarters. Through use of a national match, as in the 2000 census evaluation, the Census Bureau will be able to estimate the number of duplicates both between those in the group quarters population and those in the nongroup quarters population and the number of duplicates entirely within the group quarters population (see Mule, 2002, for the rate of duplication for various types of group quarters in the 2000 census). However, the number of omissions for group quarters residents will not be measured in 2010, nor will the number of group quarters and their residents who are counted in the wrong place.

Given the variety of ways that group quarters are enumerated, and given the various types of group quarters, coverage evaluation methods will probably need to be tailored to the specific type. We are unclear about the best way to proceed, but it is crucial that the Census Bureau find a reliable way to measure the coverage error for this group, which has been unmeasured for two censuses, going on a third. It is likely that there are sources of information, which if retained, could be used to help evaluate various proposals for measuring coverage error for group quarters residents in 2020.

What is needed is that the list of residents as of Census Day for a sample of group quarters be retained, and for this sample to be drawn independently of the Census Bureau's list of group quarters. Creating such a list probably differs depending on the type of group quarters. One would take the list of residents as the ground truth, and determine whether the residents had been included in the census and at which location. These are ideas are very preliminary, and we hope to revisit this issue prior to issuing our final report. (This general topic was item A.4 on the Census Bureau's list.)

> **Recommendation 7: The Census Bureau should collect sufficient data in 2010 to support the evaluation of potential methods for assessing the omission rate of group quarters residents and the rate of locating group quarters in the wrong census geography. This is a step toward the goal of improving the accuracy of group quarters data.**

### 3–A.6    Training of Field Enumerators

The 2010 census will be the first in which handheld computing devices are used. They will be used in the national block canvass to collect information on addresses to improve the MAF, and they will also be used for nonresponse follow-up and for coverage follow-up. While the implementation of handheld computing devices was tested in the 2006 census test and will be tested further in the 2008 dress rehearsal, there remain concerns as to how successful training will be and whether some enumerators will find the devices too difficult to comfortably learn to use in the five days allotted to training. Given that it will be extremely likely that such devices will again be used to collect information in 2020 (and in other household surveys intercensally), it would be useful to collect information on who quit, and why they quit, during the training for field enumeration work, who quit and why they quit during fieldwork, and the effectiveness of the remaining enumerators using the devices. In addition, any characteristics information that would be available from their employment applications should be retained as potential predictors for the above. Finally, the Census Bureau should undertake some exit interviews of those leaving training early and those quitting fieldwork early to determine whether their actions were due to discomfort with the handheld devices. This might provide some information either about training that would be useful in adjusting the training used in 2020, or about the ease of use of the devices or about hiring criteria. (This issue is consistent with item D.3 on the Census Bureau's list.)

### 3–B    A GENERAL APPROACH TO CENSUS EVALUATION

The panel also has some general advice on selecting and structuring census evaluations. As mentioned above, the evaluations in 2000 were not as useful as they could have been in providing detailed assessments as to the types of individuals, housing units, households, and areas for which various census processes performed more or less effectively. This is not to say that an assessment of general functioning is not important, since processes that experienced delays or other problems are certainly candidates for improvement. However, evaluations focused on general functioning do not usually provide as much help in pointing the way toward improving census processes as analyses for subdomains or analyses that examine the interactions of various factors. Since the costs of such analyses are modest, we strongly support the use of evaluations for this purpose. This issue was addressed in *The 2000 Census: Counting Under Adversity*, which makes the following recommendation, which this panel supports (National Research Council, 2004a:Rec. 9.2):

> The Census Bureau should materially strengthen the evaluation [including experimentation] component of the 2010 census, including the ongoing testing program for 2010. Plans for census evaluation studies should include clear articulation of each study's relevance to overall census goals and objectives; connections between research findings and operational decisions should be made clear. The evaluation studies must be less focused on documentation and accounting of processes and more on exploratory and confirmatory research while still clearly documenting data quality.
>
> To this end, the 2010 census evaluation program should:
> 1. identify important areas for evaluations (in terms of both 2010 census operations and 2020 census planning) to meet the needs of users and census planners and set evaluation priorities accordingly;
> 2. design and document data collection and processing systems so that information can be readily extracted to support timely, useful evaluation studies;
> 3. focus on analysis, including use of graphical and other exploratory data analysis tools to identify patterns (e.g., mail return rates, imputation rates) for geographic areas and population groups that may suggest reasons for variations in data quality and ways to improve quality (such tools could also be useful in managing census operations);
> 4. consider ways to incorporate real-time evaluation during the conduct of the census;
> 5. give priority to development of technical staff resources for research, testing, and evaluation; and
> 6. share preliminary analyses with outside researchers for critical assessment and feedback.

Item (3) is particularly important, in stressing the need for analysis, not just summaries of the (national) functioning of various census processes.

We think that evaluations should attempt to answer two types of questions. First, evaluations should be used to support or reject leading hypotheses about the effects on census costs or data quality of various census processes. Some of these hypotheses would be related to the list of topics and questions that were provided to the panel, but more quantitatively expressed. For example, such a hypothesis might be that bilingual questionnaire delivery will increase mail response rates in the areas in which it is currently provided in comparison with not using this technique. To address this question, assuming that targeting of mail questionnaires to all areas with a large primarily Spanish-speaking population is used, one might compare the mail response for areas just above the threshold that initiates this process to those just below. While certainly not as reliable or useful as a true experiment, analyses such as these could provide useful evidence for the assessment of various component processes without any impact on the functioning of the 2010 census.

Second, comprehensive data from the 2010 census, its management information systems, the 2010 census coverage measurement program, and contextual data from the American Community Survey and from administrative records need to be saved in an accessible form to support more exploratory analysis of census processes, including graphical displays. Each census surprises analysts with unforeseen problems, such as the large number of duplicate addresses in the 2000 census, and it is important to look for such unanticipated patterns so that their causes can be investigated. Standard exploratory models should be helpful in identifying these unanticipated patterns. Of course, any findings would need to be corroborated with additional testing and evaluation.

## 3–C  INITIAL CONSIDERATIONS REGARDING A GENERAL APPROACH TO CENSUS RESEARCH

The Census Bureau has a long and justifiably proud history of producing important research findings in areas relevant to decennial census methodology. However, the panel is concerned that in more recent times research has not played as important a role in census redesign as it has in the past. Furthermore, there is the related concern that research is not receiving the priority and support it needs to provide the results needed to help guide census redesign. We give four examples to explain this concern.

First, research in areas in which the results were relatively clear has been unnecessarily repeated. An example is the testing of the benefits from the use of a targeted replacement questionnaire, which was examined during the 1990s and also in 2003. The increased response resulting from the use of a targeted replacement questionnaire was relatively clear based on research carried out in the 1970s by Dillman (1978). In 1992 the Census Bureau carried out the Simplified Questionnaire Test (SQT), which examined the use of a blanket replacement questionnaire. Dillman et al. (1993a,b) describe the Implementation Test (IT), also carried out in 1992, which attempted to determine the contribution of each part of the mailing strategy toward improving response. As a result of the SQT and the IT, Dillman et al. (1993a,b) estimated that the second mailing would increase response by 10.4 percent. Subsequently, the Census Bureau also carried out two studies investigating the impact of a second mailing in hard-to-count areas. Dillman et al. (1994) showed that a second mailing added 10.5 percent to the response rate. Given the findings of this research, it is unclear why there was a need to examine the benefits from the use of a replacement questionnaire in the 2003 census test (National Research Council, 2003).

Second, areas in which research has demonstrated clear preferences have been ignored in subsequent research projects, when, for example, the previously preferred alternative was not included as a control (see National Research Council, 2006:Box 5-3). Furthermore, there are some basic questions that never get sufficient priority because they are by their nature long-term questions. The best way to represent residence rules is an obvious example. Finally, the analysis of a test census is often not completed in time for the design of the next test census, therefore preventing the continuous development of research questions.

The Census Bureau needs to develop a long-term plan for obtaining knowledge about census methodology in which the research undertaken at each point in time fully reflects what has already been learned so that the research program is truly cumulative. This research should be firmly grounded in the priorities of improving data quality and reducing census costs. Research continuity is important not only to reduce redundancy and to ensure that findings are known and utilized, but also because there are a number of issues that come up repeatedly over many censuses that are inherently complex and therefore benefit from testing in a variety of circumstances in an organized way, as unaffected as possible by the census cycle. These issues therefore need a program of sustained research that extends over more than a single decennial cycle. Also, giving people more freedom to pursue research issues may reduce turnover in talented staff.

Finally, given the fielding of the American Community Survey, there is now a real opportunity for research on census and survey methodology to be more continuous. These preliminary considerations will be greatly amplified by the panel in its subsequent activities. In the meantime, we make the following recommendation as an indication of the overall theme for which the panel anticipates developing a more refined and detailed message in later reports.

> *Recommendation 8:* **The Census Bureau should support a dedicated research program in census and survey methodology, whose work is relatively unaffected by the cycle of the decen-**

nial census. In that way, a body of research findings can be generated that will be relevant to more than one census and to other household surveys.

For example, the Census Bureau can determine what is the best way to improve response to a mailed questionnaire through use of mailing materials and reminders, or what is the best way using a paper questionnaire or the Internet to query people as to their race and ethnicity, or what is the best way using a paper questionnaire or the Internet to query people as to the residents of a household. The objective will be to learn things whose truth could be applied in many survey settings and to create an environment of continual learning, and then document that learning, to create the best state-of-the-art information on which to base future decisions. When an answer to some issue is determined, that information can be applied to a variety of censuses and surveys, possibly with modest adaptations for the situation at hand. This is preferable to a situation in which every survey and census instrument is viewed as idiosyncratic and therefore in need of its own research projects. However, one complication of developing a continuous research program on censuses and surveys is the different environments that censuses and surveys of various kinds represent. We hope to have more to say on how to deal with this in our final report.

As pointed out by the Panel on Residence Rules in the Decennial Census, "Sustained research needs to attain a place of prominence in the Bureau's priorities. The Bureau needs to view a steady stream of research as an investment in its own infrastructure that—in due course—will permit more accurate counting, improve the quality of census operations, and otherwise improve its products for the country" (National Research Council, 2006:271). A major objective of the remainder of the panel's work will be to provide more specifics on how such a research group could develop and carry out a research program in various areas and overall, and how they would make use of the various venues and techniques for research, testing, experimentation, and evaluation.

# – 4 –

# Considerations for the 2010 Census

In carrying out our primary charge regarding the selection of experiments and evaluations for the 2010 census, the panel inevitably had to consider plans for the conduct of the census itself. Moreover, the conduct of every census inevitably affects the Census Bureau's overall research program for the decennial censuses. Thus, in this chapter the panel presents three recommendations concerning some census operations with a view to their contributions to improvement to census methodology. Although we understand that the design of the 2010 census is relatively fixed, we hope that the material in this chapter may still be of use to the Bureau.

## 4–A  TECHNOLOGY

The Census Bureau will be using more technology in the 2010 census than in previous censuses, and this has raised some concerns that the panel would like to see addressed in the final plans for 2010. The concerns involve the functioning of the handheld computing devices to collect field enumeration data and the operation of the management information system for the 2010 census. By management information system is meant the various software systems that manage and monitor, somewhat interactively, the mailout-mailback process, nonresponse follow-up, field enumerator hiring and firing and compensation, questionnaire data capture, and other major census processes. We don't know the full extent to which these systems need to interoperate, but at least some modest degree of interaction is required, for example between the Master Address File (MAF)–TIGER system and the handheld devices in providing electronic maps for the handheld devices to display. The two primary concerns are whether the transmission of data using the handheld computing devices could be compromised in some manner (or could be lost unintentionally through mistakes and technological problems) and whether the needed interoperability of the components of the management information system could be hampered either by the adapting of software or the acquisition of newer software releases for the various components of the system between the dress rehearsal and the 2010 census.

With respect to the security of the transmissions of the handheld computing devices, the motivation to do harm to the census counts may be relatively modest given the lack of a financial incentive, and this may result in less chance for a security breach. However, this argument is not compelling. Furthermore, not only is there interest in reducing the opportunity for a security breach, there is also the matter of being able to assure census data users that the counts are valid. To accomplish this, the Census Bureau should carry out an independent validation and verification of the functioning of the handheld devices. This could be accomplished in the following ways, either in the 2008 dress rehearsal or in the 2010 census:

1. Establish a dual recording stream for all data from mail-in, telephone, or handheld devices: one file to go to the contractors and one to be retained by the Census Bureau. In the event of catastrophic failure by a contractor or a serious challenge to the results, it will be important to have all the raw data in the hands of the Census Bureau.

2. It is practical to develop simple programs, written and run by Census Bureau personnel, that will search large data files for patterns of interest. In this way, unexpected or curious results can be efficiently discovered and checked, and this can contribute to the validation and verification effort.

3. Related to points (1) and (2), the Census Bureau should develop quantitative validation metrics, a priori, to check for data set self-consistency and comparison of redundant data.

Other important general operational measures that we recommend for the 2010 census, either to determine whether any security breaches have occurred or to prove that the 2010 Census was secure (and which are probably already carried out), include:

- Retention of an archive of all raw data with date and time stamps. In the event of serious software failure, it would be important to be able to "replay the census" from these raw data.

- Use, by the Census Bureau and contractors, of dedicated processing systems that run no other applications and have highly secured network connections and secure accounts.

- Use of periodic system checkpoints to monitor and analyze software systems for intrusions or unauthorized manipulations of data.

- Strict control over handheld devices, including their inventory, individual device identification, and permission to operate (turn them on, turn them off, enable data transfer, disable data transfer, etc.).
- Use of methods to prevent and detect bogus data streams, including data that impersonate handheld devices.

With respect to concerns about configuration control of the management information system of the 2010 census, the processing history of the dress rehearsal could be retained and the software systems intended for use in 2010 could be used to "replay" the dress rehearsal soon before the 2010 census to identify any systems that fail to interoperate. That is, assuming that the management information system for the dress rehearsal functions well, saving the processing history would then provide a means for determining whether modifications or updates of components of the management information system between 2008 and 2010 had raised any interoperability problems. (This is referred to as regression testing.) In addition, all information system errors encountered during the dress rehearsal should be captured in a form that allows them to be used during the software development work between the dress rehearsal and the start of the 2010 census.

> *Recommendation 9:* **The Census Bureau should use dual-recording systems, quantitative validation metrics, dedicated processing systems, periodic system checkpoints, strict control over handheld devices, and related techniques to ensure and then verify the accuracy of the data collected from handheld computing devices.**

> *Recommendation 10:* **The Census Bureau should provide for a check to ensure that the subsystems of the management information system used in 2010 have no interoperability problems.**

## 4–B    DATA RETENTION BY CENSUS CONTRACTORS

Given the very successful use of contractors to carry out several decennial census processes in the 2000 census, it is expected that the use of contractors will be expanded in 2010. The component processes that will be contracted out in 2010 include (1) the decennial response integration system (DRIS), which involves systems management of the process of questionnaire response and data capture; (2) the automation of field data collection (FDCA); (3) the data access and dissemination system II (DADS II); (4) the 2010 census communications campaign; and (5) the printing contract. The fact that these systems will be operated by contractors raises an additional complication. Any data collected as part of developmental or operational testing of these systems prior to their use in 2010, as well as any data collected in monitoring the operations of these systems while in use in 2010, may be viewed as proprietary. This would limit the Census Bureau's ability to assess the performance of these systems in looking toward 2020. While the contractors themselves may issue their own evaluation studies, this is insufficient given that contractors have a bias in evaluating their own systems. We assume that contractual agreements about the sharing of such data, if they have not already been provided for, are now too late (especially for developmental testing results). In that event, the Census Bureau should try to develop some informal understandings of data sharing with their contractors to address this issue. If it is not too late, such data-sharing clauses should be included in contracts.

## 4–C  CENSUS ENUMERATION AS PART OF TELEPHONE QUESTIONNAIRE ASSISTANCE

The current plans regarding the use of Telephone Questionnaire Assistance (TQA)[1] are for it to function primarily as a means for assisting the public in making correct responses to the census form, in particular for complicated situations involving residence rules or responses to the race and ethnicity questions. In addition, this is a method for people to obtain assistance in filling out the census questionnaire when English is not their primary language. On occasion, this has also been a vehicle for households to provide their responses to the census questionnaire. However, this possibility was not encouraged in 2000.

For the 2010 census, we think the Census Bureau should consider making more transparent to respondents this option of collecting the information for the entire census questionnaire over the telephone once someone calls TQA. Specifically, whenever someone connects to TQA, the willingness of the operator to take the complete information, instead of just providing the specific help requested, should be made known to the caller during the initial part of the interaction. Our understanding is that this was not done in previous censuses due to the resources needed, especially the number of operators, and due to the additional procedural complications, especially of providing this opportunity for those receiving the census long form. However, given that this is a short-from-only census, we think that the need to get the information as soon as possible, when possible, should outweigh other concerns about making this option more frequently used. This could be especially important if the hourly wages of field enumerators increase

---

[1] Telephone Questionnaire Assistance was an operation used in the 2000 census in which people could call a toll-free number to get help in filling out their census questionnaire, to arrange to be sent a replacement questionnaire, to arrange to be sent a language guide, or to provide their census questionnaire information in situations in which they were not provided a census questionnaire.

substantially in 2010, since collection of such information may importantly reduce the cost of the nonresponse follow-up.

If this change is not implemented in 2010, the Census Bureau should collect sufficient information to carry out an evaluation after the census is completed as to the percentage of callers to TQA who ultimately sent back their census questionnaires to estimate the additional nonresponse follow-up costs due to the lack of collection of the entire census questionnaire over the telephone. Also, a possible experiment that should be considered is to sample the callers and ask those sampled if they would mind providing their information at that time by telephone to better estimate the additional resources required.

> *Recommendation 11:* **The Census Bureau should strongly consider, for the 2010 census, explicit encouragement of the collection of all data on the census questionnaire for people using Telephone Questionnaire Assistance. In addition, the Census Bureau should collect sufficient information to estimate the percentage of callers to Telephone Questionnaire Assistance who did not ultimately send back their census questionnaires. This would provide an estimate of the additional costs of nonresponse follow-up due to the failure to collect the entire census questionnaire for those cases. The Census Bureau should also consider carrying out an experiment whereby a sample of callers to Telephone Questionnaire Assistance are asked whether they would mind providing their full information to better estimate the additional resources required as a result of expanding Telephone Questionnaire Assistance in this way.**

In conclusion, the panel is enthusiastic about the opportunity to collaborate with the Census Bureau on its plans for selecting and designing productive experiments and evaluations in conjunction with the 2010 census and, more broadly, a more productive research program overall. The Census Bureau has a very proud history of innovation, including the development of punch card machines, the first nonmilitary application of computers, survey sampling, hot-deck imputation, FOSDIC (Film Optical Sensing Device for Input to Computers), to name a few, and we hope to help continue this important tradition.

# – A –

# The Census Bureau's Suggested Topics for Research

The following chart was provided to the panel by the Census Bureau as a partial summarization (augmented by several other reports and presentations) of their deliberations as to the research topics that should be considered for either experimentation during the 2010 census or evaluation shortly after. The leftmost column provides an identification key for each topic along with a short series of either questions or a brief discussion that defines the topic. The next block of columns provides criteria that should be used to help rank these topics, initiated by a high-medium-low ranking of the resulting importance of the topic. The criteria are anticipated impacts on cost, quality of data, whether the topic would require a new census component process, and whether it was accomplishable. Finally, the last block of columns provides information on whether the topic was better suited to 2010 or 2020 and whether a census environment was needed to assess alternatives to current census processes.

| Topics and Questions | Criteria | | | | | | Considerations | | |
|---|---|---|---|---|---|---|---|---|---|
| | Rank 1 = H 2 = M 3 = L | Cost (Big Payoff) | Quality | New to Census | Accomplishable | Other Criteria | For 2010? | For 2020? | Census Environment Required |
| **A. Coverage Measurement** | | | | | | | | | |
| A.1 Census Coverage Measurement (CCM) is the program that will answer the question: How accurate was the coverage of the population? | 1 | Yes | Yes | No | Yes | | Yes | Yes | Yes |
| A.2 How effective is the CCM interview and subsequent processing in determining the members of the household at each housing unit on CCM interview day and the usual residence of each household member on Census Day? When there are errors in determining household membership and usual residence, what are the causes and what are the possible remedies? What are the effects of recall errors and reporting errors on the CCM interview? | 1 | No | Yes | Yes | Yes | | Yes | Yes | Yes. (Note: Some things can be done outside the census.) |
| A.3 Can we start to learn if comparing the history of census operations with the CCM results in the sample blocks can help us explain how and when errors occur, and also suggest potential remedies? | 1 | No | Yes | No | Yes, as a feasibility study in small number of blocks in the CCM sample. | | No | Yes | Maybe |

| Topics and Questions | Rank 1 = H 2 = M 3 = L | Criteria | | | | | Considerations | | |
|---|---|---|---|---|---|---|---|---|---|
| | | Cost (Big Payoff) | Quality | New to Census | Accomplishable | Other Criteria | For 2010? | For 2020? | Census Environment Required |
| A.4 Our knowledge about Group Quarters (GQ) coverage is very limited. Efforts to estimate GQ coverage in 1980 and 1990 were limited (in both scope and success), and the GQ population was out-of-scope for the 2000 Accuracy and Coverage Evaluation (A.C.E). Since the GQ population will also be out-of-scope for CCM in 2010, we need to consider that the problems are likely to be different for different types of GQs (e.g., college dorms vs. nursing homes vs. migrant farm worker camps). | 1 | No | Yes | No | Yes | Always a problem | No | Yes | Yes |
| A.5 How can we develop a standard of comparison for household membership on CCM interview day and usual residence on Census Day? What are the effects of recall errors and reporting errors on the CCM interview? Candidate methods for developing the standard include ethnographic studies matched to the census and CCM interviews, respondent debriefings following a CCM interview, and an in-depth Living Situation Survey. | 2 | No | Yes | Yes | Yes, with limitations (scope must be very limited, and any "standard" cannot be expected to get exact truth). | | Yes | Yes | Yes, Ideally |
| A.6 Can administrative records augment CCM fieldwork from telephone follow-up to reduce cost and improve CCM data quality? For example, administrative records information may aid in confirming which enumerations linked in the computerized search for duplicates are the same person when the determination cannot be made in the field. An evaluation of A.C.E. Revision II estimates of duplication in Census 2000 using administrative records information demonstrated potential for improving CCM data quality in this manner. | 2 | Yes | Yes, unclear | N/A | Yes | | No | Yes | Yes |

| Topics and Questions | Rank 1 = H 2 = M 3 = L | Criteria Cost (Big Payoff) | Quality | New to Census | Accomplishable | Other Criteria | Considerations For 2010? | For 2020? | Census Environment Required |
|---|---|---|---|---|---|---|---|---|---|
| **B. Race and Hispanic Origin** | | | | | | | | | |
| B.1 Evaluate alternative race and Hispanic origin questions to include (1) double-banking of response categories and shared write-in spaces, (2) modified examples to follow the Advisory Panel recommendations, (3) separate evaluation of the features of 2005 National Census Test (NCT) panel 6, to better understand how each influences Hispanic and race reporting, and to inform future decisions, (4) modified Hispanic question that allows multiple Hispanic reporting (Y/N, yes multiple types). The latter must be tested in both Nonresponse Followup (NRFU) and the mailout. Samples must adequately represent small groups. Re-interview is needed to assess data quality. | 1 | No | Yes | No | Yes | | No | Yes | Desirable especially for small groups |
| B.2 Develop a combined race and Hispanic origin question. | 3 | No | Yes | No | Yes | | Yes | Yes | Yes |
| B.3 Conduct Research to support rules for editing problematic race and Hispanic origin responses (e.g., Y/N responses to Hispanic origin). A goal is to better understand respondent intent of write-in entries in the presence of, and in the absence of, marking checkboxes. | 1 | No | Yes | No | Yes | | No | Yes | Yes |
| **C. Coverage Improvement** | | | | | | | | | |
| **Address List Development** | | | | | | | | | |
| C.1 How accurate was the final address list? | 1 | No | Yes | No | Yes | | Yes | Yes | Yes |
| C.2 How should we deal with updating the address frame coming out of the 2010 Census, so that we can avoid a large and expensive address canvassing operation in the future, or so that the operation could be conducted at a much reduced cost? | 1 | Yes | Yes | Yes | See Below (C.2.a) | | Yes | Yes | Not sure |
| C.2.a How can the quality of the address frame be improved with a more scientific extract process? | 1 | Yes | Yes | Yes | Yes, we need evaluations to demonstrate that this can be done. | | Yes | Yes | Not sure |

| Topics and Questions | Rank 1 = H 2 = M 3 = L | Cost (Big Payoff) | Quality | New to Census | Accomplishable | Other Criteria | For 2010? | For 2020? | Census Environment Required |
|---|---|---|---|---|---|---|---|---|---|
| | | | | Criteria | | | | Considerations | |
| C.2.b How can we use additional information (like the Delivery Sequence File in rural areas, American Community Survey (ACS) Time of Interview data, Carrier Route data, and the National Change Of Address file) to improve address list maintenance? | 1 | Yes | Yes | Yes | Yes, we need evaluations to demonstrate that this can be done. | | Yes | Yes | Not sure |
| C.3 Can we target Address Canvassing activities better? | 1 | Yes | No | No | Yes, there are concerns about the ability to reliably match persons with common names and across long distances | Politically acceptable – Can we convince stakeholders that we don't need to do an address canvassing in their jurisdiction, while we do need to do it in the neighboring jurisdiction. | Yes | Yes | No; need 2010 data |
| C.4 How accurate were the data collected in Address Canvassing? How can we improve Address Canvassing quality? | 1 | No | Yes | No | Yes | | Yes | Yes | Yes |
| C.4.a How well does automated Global Positioning System (GPS) collection work in terms of completeness and accuracy of GPS coordinate data? | 1 | No | Yes | No | Yes | | Yes | Yes | Yes |
| C.4.b How can we improve GPS collection—increase human intervention, improve automated collection, both? | 1 | No | Yes | No | Yes | | Yes | Yes | Yes |
| C.5 How can we improve address list maintenance, operational procedures, and enumeration of small multi-unit structures (2-10 units)? | 2 | No | Yes | No | Yes | | Yes | Yes | Yes |

| Topics and Questions | Rank 1 = H 2 = M 3 = L | Criteria | | | | | Considerations | | |
|---|---|---|---|---|---|---|---|---|---|
| | | Cost (Big Payoff) | Quality | New to Census | Accomplishable | Other Criteria | For 2010? | For 2020? | Census Environment Required |
| **Administrative Records** | | | | | | | | | |
| C.6 How can we avoid the need for followup and use administrative records to: a. Identify coverage problems? b. Identify and classify duplicates? c. Resolve potential coverage problems identified by the coverage probes. | 1 | Yes | Yes | Yes | Yes | Privacy concerns and issues with file access. Stakeholders have expressed reservations about the use of administrative records | Not sure | Yes | Yes |

| Topics and Questions | Rank 1 = H 2 = M 3 = L | Cost (Big Payoff) | Quality | New to Census | Accomplishable | Other Criteria | For 2010? | For 2020? | Census Environment Required |
|---|---|---|---|---|---|---|---|---|---|
| **Coverage Followup (CFU)** | | | | | | | | | |
| C.7 Does Coverage Followup actually work? <br> a. How effective is it? <br> b. Is CFU effectively identifying omissions? <br> c. Is it introducing bias? <br> d. How do recall and reporting errors affect its determination of residency, and hence erroneous enumerations (EEs)? <br> e. How can we afford to follow up on more coverage improvement cases? <br> f. Is the expense of CFU worth the coverage gain? <br> g. Can certain categories of response to coverage questions be automatically coded, or field coded by interviewers to reduce follow up workload? . . . . <br> h. What recall and reporting problems affect CFU's ability to identify: missed people the respondent had in mind when filling out the undercount question? Are enumerators screening out people who are eligible to be listed? <br> i. How to optimize which cases are coded for CFU? | 1 | Yes | Yes | Yes | Yes | | Yes | Yes | Yes |
| C.8 Develop and experimentally evaluate alternative designs for coverage followup instruments. Alternative methodologies might involve dependent questions; self-response by all relevant household members; immediate follow-up; and other methodological improvements to facilitate recall and reporting in CFU (i.e., conduct an integrated experiment). | 1 | Yes | Yes | Yes | Yes | | Yes | Yes | Yes |

| Topics and Questions | Rank 1 = H 2 = M 3 = L | Criteria | | | | | Considerations | | |
|---|---|---|---|---|---|---|---|---|---|
| | | Cost (Big Payoff) | Quality | New to Census | Accomplishable | Other Criteria | For 2010? | For 2020? | Census Environment Required |
| C.9 To what extent did nationwide person matching improve the identification and removal of duplicates of housing units and persons in the census? In particular, what improvements can be made in the identification and removal of census duplicates of persons across some distance given the challenges created by chance agreements of names and birth dates? | 1 | Yes | Yes | Yes | Yes there are concerns about the ability to reliably match persons with common names and across long distances. | | Yes | Yes | Yes |
| C.10 Develop and experimentally evaluate alternative designs of the undercount (and overcount?) questions in the mail form to effectively identify census coverage errors for follow-up. Variations might include format (open vs. closed) and wording of questions and response categories, and placement in the form. | 1 | Yes | Yes | Yes | Yes | | Yes | Yes | Yes |
| **Residency Rules/Questionnaire Design** | | | | | | | | | |
| C.11 Implement and experimentally evaluate alternative residence rules and presentation of roster instructions in paper and other modes, including the National Academy of Sciences (NAS) recommendations to ask a sufficient number of residence questions to determine residence, and to obtain alternative addresses. Panels would be included in an alternative questionnaire experiment (AQE) and would require a coverage re-interview. Cognitive testing is needed for development, along with research on respondents' reading behavior and use of flashcards or other ways of presenting instructions. Alternative approaches might include: a. de facto approach b. Worksheet approach c. Alternate address elsewhere | 2 | No | Yes | No | Yes | | No | Yes | Yes, for some aspects |

| Topics and Questions | Rank 1 = H 2 = M 3 = L | Criteria | | | | | Considerations | | |
|---|---|---|---|---|---|---|---|---|---|
| | | Cost (Big Payoff) | Quality | New to Census | Accomplishable | Other Criteria | For 2010? | For 2020? | Census Environment Required |
| **Be Counted** | | | | | | | | | |
| C.12 What effects does the Be Counted Program have in filling gaps in coverage? a. Is it worth it? b. Is including it better than trying to ensure people are counted in other ways? c. Does it introduce coverage errors? | 3 | No | Yes | No | Yes | Public perception —of missing people or not giving them an opportunity to be enumerated in the census if not for the Be Counted program. | Yes | Yes | Yes |
| **General** | | | | | | | | | |
| C.13 How accurate was vacancy/occupancy status of housing units in Census? Are there ways to improve accuracy? | 1 | No | Yes | No | Yes | | Yes | Yes | Yes |
| C.14 Through ethnographic research, can we learn more about American Indian and Alaska Native households, Hispanic households, and immigrant communities that might result in different methods for enumeration? This research could provide insight into CFU and Census Coverage Measurement (CCM) to understand deficiencies. | 2 | No | Yes | No | Yes | | No | Yes | Yes |

| Topics and Questions | Rank 1 = H 2 = M 3 = L | Criteria | | | | | Considerations | | |
| --- | --- | --- | --- | --- | --- | --- | --- | --- | --- |
| | | Cost (Big Payoff) | Quality | New to Census | Accomplishable | Other Criteria | For 2010? | For 2020? | Census Environment Required |
| **D. Field Activities** | | | | | | | | | |
| **Automation** | | | | | | | | | |
| D.1 What was the impact of adding expanded automation to field data collection for Address Canvassing, Nonresponse Followup (NRFU) and Census Coverage Measurement - Personal Interview (CCM-PI)? Did we gain in efficiency? Did we see cost savings? Did automation contribute to operational improvements? Should we use the hand-held computers (HHCs) in operations other than Address Canvassing, NRFU and CCM-PI in 2020(e.g., U/E)? | 1 | Yes | Yes | Yes | Yes | | Yes | Yes | Yes |
| D.2 What was the impact on field staff of using HHCs to conduct field data collection operations? Did using HHCs help us to improve the effectiveness and efficiency of field staff? Did using the HHC help us improve the productivity of field workers? What impact did the HHC have on field staff training? | 2 | Yes | Yes | Yes | Yes | | Yes | Yes | No |
| **Training** | | | | | | | | | |
| D.3 How can enumerator training be improved? a. Can we make enumerator training more efficient/effective through the redesign of enumerator training materials and job aids? b. Can we make enumerator training more effective by expanding the use of technology-based training? | 1 | Yes | Yes | Yes | Yes | | Yes | Yes | |

| Topics and Questions | Rank 1 = H 2 = M 3 = L | Criteria | | | | | Considerations | | |
| --- | --- | --- | --- | --- | --- | --- | --- | --- | --- |
| | | Cost (Big Payoff) | Quality | New to Census | Accomplishable | Other Criteria | For 2010? | For 2020? | Census Environment Required |
| D.4 How can we better prepare enumerators to ensure they are effective and efficient in their job? What is the optimal contact strategy for NRFU? How many contacts should we make for NRFU? (Requires experimental design). | 1 | Yes | TBD | No | This research requires an experiment design with different contact strategy in different locations. Implementing such a design in the census environment may not be practical or worthwhile | | Yes | Yes | Yes |
| D.5 How can enumerator training be improved to reduce/minimize errors that may be introduced by interviewers? The focus is on interviewers' contributions to coverage errors (in NRFU, CFU, and CCM-PI) in particular, but also errors in other short form items. Possible research approaches might include: an interviewer variance study, in which interviewer assignments are randomized, or assigning a sample of mail returns to NRFU enumerators for re-interview. | 2 | No | Yes | Yes | There are studies that could be done to determine interviewer contribution to error. However, we are not sure if training could be improved or changed to address those contributions. | | Yes | Yes | Yes |

| Topics and Questions | Rank 1=H 2=M 3=L | Criteria | | | | | Considerations | | |
|---|---|---|---|---|---|---|---|---|---|
| | | Cost (Big Payoff) | Quality | New to Census | Accomplishable | Other Criteria | For 2010? | For 2020? | Census Environment Required |
| **Quality Control (QC)** | | | | | | | | | |
| D.6 How can Global Positioning System (GPS) Technology be used as a QC tool for field work, e.g., to identify curbstoning or inefficient field work? | 1 | Yes | Yes | Yes | Yes | Privacy concerns—GPS Tracking of employees | No | Yes | Yes |
| D.7 How can the QC design for field operations be improved to be more effective/efficient?<br>a. How much does the QC improve the quality of the census operations? Does the QC have a high probability of identifying data falsification and/or violation of procedures?<br>b. Is there an efficient way to verify the QC work? Is it worth it to verify the QC work? | 2 | No | Yes | No | Yes | | Yes | Yes | Yes |
| D.8 Can a batch level approach to re-interview sampling improve efficiency and/or effectiveness of field re-interview operations? | 3 | Yes | Yes | No | Yes | | No | Yes | Yes |
| **E. Language** | | | | | | | | | |
| E.1 Can an alternative design for the bilingual English/Spanish questionnaire result in improved data? | 1 | Yes | Yes | Yes | Yes | Backlash; referring to lowered mail response from non-Spanish-speaking populations | Yes | Yes | Yes |
| E.2 Is there a better or more efficient way to stratify the mailing of the bilingual forms? | 2 | Low | Low | Yes | Yes | | No | Yes | No |

| Topics and Questions | Rank 1 = H 2 = M 3 = L | Criteria | | | | | Considerations | | |
|---|---|---|---|---|---|---|---|---|---|
| | | Cost (Big Payoff) | Quality | New to Census | Accomplishable | Other Criteria | For 2010? | For 2020? | Census Environment Required |
| E.3 What information can systematic observations yield about how census enumerators are obtaining information from households with little or no understanding of English? Are there changes we can or should make to our methodologies and practices to improve these interviews? | 2 | No | Yes | No | Yes, but may present complex stratification issues | | Yes | Yes | Yes |
| E.4 Can we obtain better mail response and/or higher quality data by mailing a Language Assistance Guide Booklet that depicts the questionnaire in Five (5) languages? | 1 | Yes | Yes | Yes | Yes | | No | Yes | Yes |
| **F. Mode Effects** | | | | | | | | | |
| F.1 What is the magnitude of the effects of mode on responses to 2010 census questions? This research would compare the 2010 mail mode content to the adaptation of the specific content items for other modes used in the 2010 census. An example would be comparing the 2010 mail form relationship question, which shows all 14 categories, to the proposed 2010 telephone-adapted version, which asks an open-ended question. In general, this study would examine response distributions (or reliability and other data quality measures) for the 2010 mail items compared to the adapted versions used in 2010 for other modes; comparable random samples would be ideal to avoid self-selection confounds. | 1 | No | Yes | No | Designing an experiment that reduces/ eliminates the self-selection bias is complex and may not be feasible from a field/budget/ schedule standpoint. | | Yes | Yes | Yes |

| Topics and Questions | Rank 1 = H 2 = M 3 = L | Criteria | | | | | Considerations | | |
| --- | --- | --- | --- | --- | --- | --- | --- | --- | --- |
| | | Cost (Big Payoff) | Quality | New to Census | Accomplishable | Other Criteria | For 2010? | For 2020? | Census Environment Required |
| F.2 What are the effects of mode for alternative adaptations of the 2010 census questions in non-mail modes? This research would compare the 2010 mail mode content to alternative adaptations of the specific content items for other modes. These alternative versions for the non-mail modes are adaptations, which show promise in terms of providing comparable data to the mail form, but were not used in the 2010 census. We would examine response distributions (or reliability and other data quality measures) for the 2010 mail items compared to the alternative adapted versions for other modes; comparable random samples would be ideal to avoid self-selection confounds. | 2 | No | Yes | No | Designing an experiment that reduces/ eliminates the self-selection bias is complex and may not be feasible from a field/budget/ schedule standpoint. | | Yes | Yes | Yes |
| **G. Content** | | | | | | | | | |
| G.1 What are the combined effects on the data of all questionnaire changes made in the 2010 mail questionnaire? | 1 | No | Yes | No | Yes | | Yes | Yes | Yes |
| G.2 What are the consistency and reliability of reporting in the 2010 census? | 1 | No | Yes | No | Yes | | Yes | Yes | Yes |
| G.3 How well do questions perform in interviews? | 2 | No | Yes | No | Yes | | Yes | Yes | Yes |
| G.4 How comparable are Census 2010 data and American Community Survey (ACS) data? | 1 | No | Yes | Yes | Yes | | Yes | Yes | Census data required but not environment |
| G.5 Do current methods for identifying the householder/Person 1 perform well in all modes? If not, can improved method(s) be developed? | 2 | No | Yes, possibly | No | Yes | | No | Yes | No |

| Topics and Questions | Rank 1 = H 2 = M 3 = L | Criteria | | | | | Considerations | | |
|---|---|---|---|---|---|---|---|---|---|
| | | Cost (Big Payoff) | Quality | New to Census | Accomplishable | Other Criteria | For 2010? | For 2020? | Census Environment Required |
| **H. Self-Response Options** | | | | | | | | | |
| H.1 How can we improve alternatives for increasing mail response? An experiment for 2006 showed that a deadline plus delayed mailing of questionnaires improved the mail response rate by two percentage points? This should be replicated in the 2010 Census as an experiment to see if results hold up in a census environment, and to get good data on timing of returns under a deadline. If design permits, effect of deadline messaging and compressed schedule could be teased out. | 1 | Yes | Yes | No | Yes | | No | Yes | Yes |
| H.2 Experiment testing an additional contact reminder after replacement questionnaire. This contact would contain stronger language, relative to the reminder postcard and replacement questionnaire, indicating that failure to comply would mean inclusion in the Nonresponse Followup (NRFU) workload (more expense, etc). Different types of contacts could be tested such as postcard, full size letter, phone message, etc. | 1 | Yes | Yes | No | Yes | | No | Yes | Census environment is optimal but at least site test is required to utilize NRFU message. |
| **I. Special Places/Group Quarters** | | | | | | | | | |
| I.1 Did the revised Group Quarters (GQ) definitions improve the identification and classification of GQs (GQs versus housing units, and by type)? | 1 | No | Yes | No | Yes | | Yes | Yes | Yes |
| I.2 Evaluate methods for improving GQ data collections by: 1) assessing the yield from the various sources used to update the MAF/TIGER database (MTDB), as well as assessing how the various census operations update the MTDB; 2) studying effects of allowing a Usual Home Elsewhere in more types of GQs; and 3) collecting additional information to assist with unduplication of college students. | 2 | No | Yes | Yes | Yes | | Yes | Yes | Yes |

| Topics and Questions | Criteria | | | | | | Considerations | | |
|---|---|---|---|---|---|---|---|---|---|
| | Rank 1 = H 2 = M 3 = L | Cost (Big Payoff) | Quality | New to Census | Accomplishable | Other Criteria | For 2010? | For 2020? | Census Environment Required |
| I.3 The National Academy of Sciences recommends, "The U.S. Census Bureau should participate in a comprehensive review of the consistency of content and availability of prison records. The accuracy of prisoner-reported prior addresses is uncertain, and should be assessed as a census experiment. A research and testing program, including experimentation as part of the 2010 census, should be initiated by the Census Bureau to evaluate the feasibility and cost of assigning incarcerated and institutionalized individuals, who have another address, to the other location." (National Academy of Sciences (NAS) report entitled "Once, Only Once, and in the Right Place – Residence Rules in the Decennial Census (September 2006), pp 9-10)." | 3 | No | No | Yes | Maybe—the scope of this comprehensive review of prisoner reported address information would be massive if all levels of correctional facilities (e.g., federal, state, local, and private) were included in the review. Additionally, the Census Bureau has concerns about the feasibility of actually collecting the prisoner address information. | There could be possible implications for other types of GQs. This topic has generated much discussion and varying views among census stake-holders. | Yes | Yes | Yes |
| **J. Marketing/Publicity/Paid Advertising/Partnerships** | | | | | | | | | |

| Topics and Questions | Rank 1 = H 2 = M 3 = L | Criteria | | | | | Considerations | | |
|---|---|---|---|---|---|---|---|---|---|
| | | Cost (Big Payoff) | Quality | New to Census | Accomplishable | Other Criteria | For 2010? | For 2020? | Census Environment Required |
| J.1 How effective was the communication strategy for improving response and accuracy of the census?<br>a. How do the separate components of the communications strategy contribute to the improvements (e.g., advertising, partnerships, etc.)?<br>b. How effective were the targeted messages at reaching specific audiences?<br>c. Did the communications strategy change attitudes or behavior toward and/or increase awareness of participation in the census? | 1 | Yes | Yes | No | The efforts in 2000 to evaluate the Advertising/ Marketing approach was inconclusive. Whether or not we can evaluate the 2010 approach depends on the solution put forth by the vendor and our ability to develop reliable technology. | | Yes | Yes | Yes |
| **K. Privacy** | | | | | | | | | |
| K.1 Test alternative presentation and placement of privacy messages in cover letter, etc. | 1 | Yes | Yes | No | Yes | | No | Yes | Yes |
| K.2 Monitor public concerns about privacy and confidentiality in a series of quick-turnaround surveys conducted during the census to provide U.S. Census Bureau executives with timely information about emerging concerns and issues. Data from monitoring surveys can also augment (or replace) traditional outreach evaluation surveys, which are slow and do not provide useful information on a timely basis. | 1 | No | No | Yes | Yes | Can help decision makers respond to emerging issues, crises | Yes | Yes | Yes |

## Criteria and Considerations for Assessing Proposed Research Topics and Questions

**Criteria:**

Cost (Big Payoff) – **[Yes/No]** Will results potentially lead to substantial cost savings in the 2020 Census?

Quality – **[Yes/No]** Could results conclusively measure effects on data quality?

New to Census – **[Yes/No]** Does the question address operations that are new since Census 2000, experienced significant procedural change, or experienced significant issues during Census 2000?

Accomplishable – **[Yes/No]** Will data be available to conclusively answer the question? Will there be a high demand of resources to address and answer the question? Are complex or untested methods foreseen to address and answer the question?

**Considerations:**

For 2010 – **[Yes/No]** Is this research question intended to assess an operation in the 2010 Census?

For 2020 – **[Yes/No]** Is this research question intended to assess a 2010 Census operation to inform the 2020 Census?

Census Environment Required? **[Yes/No]**

SOURCE: 2010 Census Program for Evaluations and Experiments—Appendix to Summaries of Suggested Research (planning document shared to the panel by the U.S. Census Bureau, April 13, 2007).

# – B –

# Internet Response Options in Selected Population Censuses

In this appendix, we briefly describe provisions for an online response option in past (and upcoming) censuses of population. We begin by describing the use of online response in the 2000 U.S. census and subsequent tests before describing experiences in other countries.

## B–1 THE INTERNET AND THE U.S. CENSUS

### The 2000 Census

The Internet response option was implemented in the 2000 census without the benefit of prior large-scale testing. Online response was considered for the 1998 dress rehearsal but ultimately abandoned "due to security concerns" but was revived in late 1998 by a Commerce Department directive (Whitworth, 2002:1). Due to insufficient time, online response was restricted to the 2000 census short-form questionnaire only and a single language (English). Programming of the form was kept as simple as possible for compatibility with different web browsers; JavaScript was avoided because it was deemed "unstable in some environments" (Whitworth, 2002:1). As a consequence, the online form was essentially presented as a single screen page rather than walking through separate questions in different web pages; hence, real-time editing and confirmation steps were not used, nor were skip patterns to move respondents through the questionnaire.

To access the electronic questionnaire, respondents needed to have the paper questionnaire that they received in the mail in hand. Following a link from the main census web page, they were asked to enter the 22-digit Census ID printed on the paper form's label (thus ensuring a linkage to a specific mailing address). If the 22-digit ID was confirmed as valid, then the questionnaire appeared onscreen. No publicity was given to the Internet response option.

During the time span between the opening of the online questionnaire site and the cutoff for nonresponse follow-up workload (March 3 to April 18, 2000), 89,123 submissions of Census ID numbers were made on the web site. Of these (Whitworth, 2002:5):

- 74,197 (83.3 percent) were valid Census IDs; however, only 71,333 resulted in a questionnaire submission. The other 2,864 may have been instances in which a respondent made an error entering the ID but inadvertently entered a valid number; they could have then broken off the interview and subsequently rekeyed their ID correctly. After some reconciling for unique address identifications, questionnaire data from 66,163 of the 71,133 submissions were ultimately sent on for processing; about 1,500 online submissions are unaccounted for in the Bureau's tallies, with "no apparent explanations for this discrepancy" (Whitworth, 2002:6).

- 14,926 (16.7 percent) attempts to enter a Census ID were failures. That this proportion matches the approximate 1-in-6 coverage of the census long-form sample is perhaps telling: "since [the Census Bureau] did not advertise the Internet response option, respondents would have also had no idea that long-form households were ineligible." Hence, "it is quite possible that many, if not most, of the submission failures" were attempts to use the Internet to answer a long-form questionnaire.

Although the vast majority of the Internet responses (98.4 percent) were each associated with only one ID number, there were some repeats of ID numbers: specifically, 1,090 ID numbers had to account for 2,853 responses. Most of these were incidents of 2 or 3 entries per ID and involved a pure replication of the same data; most likely, this was caused by a respondent clicking on the "Submit" button multiple times waiting for the browser page to load. The extreme case was a single ID associated with 17 entries; "many of these were on different days, and many with different data" (Whitworth, 2002:8–9). After final processing, 63,053 households representing 169,257 persons were included in the census through the Internet form.

The Census Bureau evaluation of the Internet response option in 2000 (Whitworth, 2002:17) deemed it "an operational success" and argued for further research:

Obviously, the Internet is here to stay. The software and hardware developed for this program could have handled **tens of millions** of records instead of the **tens of thousands** it did handle. It is our recommendation that future research focus not necessarily on **how** to implement the form itself, but how to promulgate the Internet form as an option and convince the public that there is sufficient data security. Future research should also focus on how to use it as a tool to increase data quality by implementing real-time data feedback and analysis.

## Response Mode and Incentive Experiment

Conducted as an experiment in the 2000 census, the Response Mode and Incentive Experiment (RMIE) gauged response rates to the 2000 census questionnaire by paper, interactive voice response (IVR, a fully automated telephone interview), or the Internet. In addition, the test considered whether the offer of an incentive (specifically, a 30-minute telephone calling card) influenced the response rates. The test (including a print of the Internet census form) is documented by Caspar (2003, 2004). The Internet usage survey component of the RMIE yielded relatively small numbers of online returns (with or without the incentive of a calling card), and some respondents noted a preference for paper. However, Caspar (2003:21) argued for further work on an online response option:

> Based on conservative assumptions and the data from RMIE, one might save between one and six million dollars in postage costs alone if between three percent and 15 percent of the sample uses the web rather than the mail survey. . . . This savings would more than offset the costs required to design, develop and maintain the web survey. Of course, the web survey would also produce savings related to reduced processing (receipt and scanning). Given this crude calculation, it is anticipated that the Internet would be cost-effective even if a relatively small proportion of respondents used it.

## The 2003 and 2005 Tests

The 2003 National Census Test was designed as a mailout-only test: no fieldwork for nonresponse follow-up was planned or conducted. The mail sample was divided into 16 panels, 7 of which tested revisions of the census questions on race and Hispanic origin and 8 of which included different packages of response modes and contact strategies (e.g., sending a replacement questionnaire or a telephone reminder call, responses by telephone or the Internet). The Census Bureau concluded that offering the option of responding by telephone or the Internet along with the mailout of a paper questionnaire neither increased nor decreased the response rate. However, attempts to "force" respondents to use either of the electronic response modes by not including a paper questionnaire resulted in lower response rates. In terms of data quality, item nonresponse rates were significantly lower for the Internet responses than for paper returns for almost all items.[1]

A second mailout-only National Census Test in 2005 made another attempt to implement the telephone and Internet response modes, having made interface improvements in both. Illustrative screens—of the respondent log-in section and the race question—from the 2005 online instrument are shown in Figure B-1. Apparently, this test performed comparably to the options used in 2003 and did not yield major gains in response.

In November 2000–January 2001, the Census Bureau also conducted a test using 10,000 addresses on an Internet response option for the American Community Survey (ACS), the replacement for the traditional census long-form questionnaire in 2010. The recent report *Using the American Community Survey: Benefits and Challenges* describes ACS methodology in greater detail (National Research Council, 2007). In brief, the sample of households selected in one month is first contacted by mail and asked to return their questionnaire by mail. If they do not respond by mail, a telephone interview is attempted in the second month; if that fails, then enumerators attempt a personal visit in the third month. The hope of an Internet response option would be to supplement mail responses in the first months so that the follow-up steps in months 2–3 need not occur. Griffin et al. (2001) found that only about 2 percent of the respondents in the experimental group used the Internet response option (compared with about 36 percent by mail). The data showed some attempts to access or partially fill out the questionnaire, but they did not result in a full online form being submitted and were not enough to explain the low response rate. Although the response was low, the quality of the resulting data (in terms of whether subsequent editing was required) was found to be slightly better in the Internet responses than the mail responses.

## Decision for 2010

An initial planning framework for the 2010 census (Angueira, 2003:3) noted among the major improvements planned for 2010 that "expanded use of Internet and telephone systems (using Interactive Voice Response) will provide new opportunities for using technology to make it easier for people to complete their questionnaire." The strategy document elaborated (Angueira, 2003:5–6):

> Fundamental to the 2010 census is expanding the ways people can be counted. Following a widespread awareness campaign, households will

---

[1]The 2003 test was summarized (albeit without specific numbers) at http://www.census. gov/procur/www/2010dris/web-briefing/dris-tel-int.html.

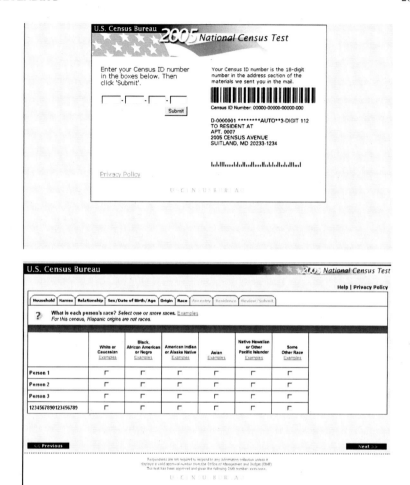

**Figure B-1**   Housing Unit ID log-in screen and race response screen,
Internet questionnaire, 2005 census test

SOURCE: http://www.census.gov/procur/www/2010dris/omb-person-based-screens.doc.

receive an advance letter in the mail before April 1, 2010. The letter will tell them about the census and the ways they can participate, using English or other language methods. . . . We will also use technology to build on this strategy by combining these mailings with Internet and telephone contacts. These technologies will provide respondents with additional options for receiving and submitting their census questionnaires. Our expectation is that we can increase the response rate even further by developing and implementing the optimal mix of contacts and response options. By taking advantage of the Internet and the telephone we can significantly increase the number of forms that move directly into data capture without needing to be scanned in a data capture center. . . .

Despite all efforts to encourage everyone to provide information, we project that we will not obtain mail, Internet or telephone IVR responses from as many as 31% of the addresses to which we deliver a questionnaire. Many of these addresses will be vacant or nonexistent, but many will be occupied. Therefore, we must still conduct a nonresponse follow-up operation. . . .

Indeed, the initial scope of work for the Census Bureau's Decennial Response Integrated System (DRIS) for 2010 included requirements to facilitate census responses by three modes: paper, telephone, and Internet. The first two objectives suggested for the DRIS solution were to "Enable the Public" to "Obtain assistance or request an English or foreign language questionnaire or language guide using the telephone or Internet" and "Complete their 2008 Dress Rehearsal and 2010 Census questionnaire via the telephone, Internet and paper."[2] The DRIS contract was awarded to Lockheed Martin in October 2005.

However, the perceived low Internet response rates in the 2003 and the 2005 tests—combined with concern over inherent risks and the lack of guaranteed major cost savings—led the Census Bureau to reverse course. The Bureau's decision not to pursue online enumeration was formalized in a July 2006 decision memorandum. Earlier, on June 6, Census Director Louis Kincannon (2006) offered the following argument in testifying before a U.S. Senate subcommittee:

We have also considered other data collection methods, including Internet data collection. Based on our research, as well as our own experience and knowledge of the experiences of other countries, we do not believe Internet data collection would significantly improve the overall response rate or reduce field data collection. The Census Bureau offers an electronic response option for the Economic Census and other economic surveys and we generally obtain high response rates. It is altogether different, however, when we consider household and population surveys and censuses. The 2003 and 2005 Census Tests offered an In-

---

[2] http://www.census.gov/procur/www/2010dris/web-briefing/dris-goals-objectives.doc.

ternet response option, and in both cases, the response rates were low, and offering an internet response option did not increase the overall response rate. We have also consulted the statistical offices of Australia, Canada, and New Zealand. Each of these countries utilized the Internet in their most recent censuses. The Internet response rate ranged from 7 to 15 percent. Each of the statistical offices indicated that it was not possible to accurately anticipate the response rate, and that ultimately using the Internet did not affect the overall response rate. Anticipating the response rate has important operational considerations. Because they were unable to accurately anticipate the Internet response rate, the other countries were unable to reduce the paper data capture operations out of concern they would not have the capacity to fully process the census responses. This would be true for the Census Bureau as well. Moreover, the Internet response option did not reduce the overall cost of data collection, and the cost for some specific activities, such as security and server capacity, increased.

We have seriously considered the lessons our colleagues have learned. We are also concerned that utilizing the Internet could jeopardize other planned improvements. At this point in the decade, efforts to develop an Internet response option would divert attention and resources from tested and planned improvements such as the second mailing—which we know can increase the overall response rate by several percentage points. It is also important to keep in mind that the 2010 Census utilizes only the short form. There are very few questions in this form, and most can be answered by checking a box.

The major risks perceived by the Census Bureau—summarized in a commissioned report from the MITRE Corporation (2007)—are as follows:[3]

- Above all, the Census Bureau is concerned that something gone awry in an Internet response option—publicity of the census site being hacked or establishment of a "phishing" site appearing to be related to the census, for example—could cause voluntary response to the census to decline. This would tax nonresponse follow-up capabilities and raise the overall cost of the census.

- The Bureau's DRIS contractor concluded that it could not provide an Internet response facility in time for testing in the 2008 dress rehearsal, so that it would have to go into the main 2010 census without a large-scale test (as happened with the 2000 census online response option).

---

[3]The MITRE report was circulated on some technology blogs in July 2007, following a Senate subcommittee hearing at which the Census Bureau restated its intent not to pursue online enumeration. At the same hearing, Sen. Tom Coburn (R-OK) issued a public "Census Challenge" for ideas to use technology to reduce the costs of the 2010 census. See, e.g., http://www.fcw.com/blogs/archives/editor/2007/07/the\_census\_inte.asp, which contained a link to the MITRE report and references an interview with a former Census Bureau official.

- A problem faced by any Internet site is a "denial of service" attack: deliberate bombardment with hits in order to shut down a site's operations.

(The MITRE evaluation also expresses concern that census data might be captured from individuals' computers through the use of spyware.)

In evaluating the Census Bureau's work on group quarters enumeration, the U.S. Department of Commerce, Office of Inspector General (2006:20–21) acknowledged the Bureau's decision not to use the Internet for main data collection in 2010. However, the review strongly suggested that the Bureau consider use of Internet methods for one traditionally hard-to-count population: college students. One reason for the selection of parts of Travis County, Texas, as a census test site in 2006 was a large college student population. Yet only 719 college student census report forms were returned during the test while expectations were that more than 6,700 should be found. In the inspector general's review, this suggested that online response options might appeal to the Internet-savvy college generation. Reacting to this recommendation, the Census Bureau reiterated its opposition to online enumeration generally.

## B–2   USE OF THE INTERNET IN FOREIGN CENSUSES

In offering guidance to member countries on the 2010 round of population and housing censuses, the United Nations Economic Commission for Europe (2006) concisely summarized the basic rationale and concerns for permitting an Internet response option; this summary is presented in Box B-1. Stopping short of recommending that countries adopt an online version, the commission observed that online response is becoming an increasingly attractive option.

In this section, we profile the use of the Internet as a response mode in selected censuses around the world, focusing almost exclusively on countries that still perform a traditional census rather than rely on a population register or other methods. Online enumeration has been performed in most of these cases; however, we also describe one census that ruled out Internet enumeration in its most recent census (Japan) and another that has not yet used the Internet in the census or in a major census test but intends to do so (United Kingdom).

One common theme to several of these profiles—particularly Canada, Australia, and New Zealand—is that the drive to allow the Internet as a response option came about through longer standing commitments to making government services electronically accessible. The Canadian "Government On-Line" initiative began in 1999, with the objective of making most government services accessible online by 2004–2005. The Canadian govern-

---

**Box B-1** United Nations Economic Commission for Europe Comments on Internet Data Collection in the 2010 Round of Censuses

Using the Internet as a collection method means that the census collection methodology will need to be self-enumeration rather than interview based. The Internet option can be incorporated into any of the traditional methods of delivering and collecting census forms (for example drop-off/pick-up, mail-out, mail back). The key factor is managing collection control operations—that is ensuring that every household and individual is counted once and once only. This requires the ability to provide each household and individual with a unique code linked to a geographic location. An added complication for those countries where forms are collected by census enumerators (rather than mailed back) is to have adequate and timely feedback to enumerators so that they can update their own collection control information so that they do not visit households that have already returned forms.

The potential level of take-up of an Internet option should be considered by assessing the proportion of the population who can access the internet from home, the proportion who use broadband services and the general use of the Internet for other business purposes (for example on-line banking, filing tax forms, shopping). The use of the Internet is likely to increase the cost of the census, at least initially. As it is not known in advance who is likely to use the Internet, there will be a need to deliver a paper form to every household including those who will subsequently use the Internet. Systems and processes that allow for Internet return of census forms will also need to be developed. These will increase costs. On the other side there are potential savings in data capture costs. However, scanning and Intelligent Character Recognition are in themselves cost efficient. Therefore, savings in data capture costs are likely to be considerable less than the costs of developing and implementing the internet system.

Security is an important consideration. Industry standard encryption (SSL128) offers two-way encryption (that is it encrypts data flowing both from and to the user's computer) and has been accepted by nearly all countries as adequate to protect the census information. Security should be a key consideration in designing the infrastructure. A physically separate infrastructure should be set up to collect the census information. Completed individual census forms should be moved behind firewalls and then into infrastructure that is completely separate from the collection infrastructure.

A downloadable on-line form requires much less infrastructure than for forms that are completed on line. However, downloadable forms require a greater level of computer literacy than on-line forms. They will not necessarily work in thousands of different computer configurations and there will be an expectation that the census agency will be able to deal with each individual problem. From the respondents' point of view, they are much more likely to prefer completing the form on-line. For these reasons it is expected that most countries will adopt on-line completion of census forms.

An electronic form offers the possibility of interactive editing to improve response quality that is not possible on a paper form. People using electronic forms have a certain level of expectation that a certain amount of guidance will be offered—at a minimum that they will be sequenced through the form and not asked questions that are not relevant to their situation. How far other editing or on-line coding is built in to the form needs to be carefully considered. Some limited studies indicate that forms returned by the Internet are of higher quality than paper forms. More work is required in this area to determine whether this is a function of the type of people using the Internet or the technology itself.

*(continued)*

---

**Box B-1**  (continued)

Providing an Internet option may contribute to improving the quality of the census by making it easier for some hard-to-enumerate groups to respond. Most countries report difficulties in enumerating young adults and people living in secured accommodation where access is restricted. Some people with disabilities will also find it easier to complete an Internet form than a paper form. These groups are also more likely to be using the Internet and, if available, this option should be promoted to these groups as a means of encouraging participation in the census.

Provision of sufficient infrastructure provides one of the major challenges for offering an Internet option. The census occurs over a relatively short period of time and affects the whole population of a country, and it is unlikely that the census agency will have adequate infrastructure to cope with the peak demands of a census. It is therefore likely that this component, at least, of the Internet solution will be outsourced. It may be necessary for collection procedures to be modified to constrain demand. For example, requiring people outside predetermined target populations/areas to contact the census agency before they can use the Internet form may be a means of restricting use of the Internet form. Census agencies need to assess how they wish to promote the use of the Internet. Promotion of the Internet option should be determined by the capacity of the service to handle the expected load and should be coordinated with the collection procedures. The public relations strategy will need to encompass assurance about security of information supplied via the Internet. Assuming that the Internet option is targeted to the whole population, the public relations strategy should encompass managing public expectations about the ability to access the site during periods of peak demand. Simple messages advising people to use the internet option at "off peak" times should be prepared and used if necessary on the census internet site itself and through the census telephone inquiry service, radio and print.

SOURCE: Excerpted from United Nations Economic Commission for Europe (2006:Paragraphs 119–125). Reprinted with the permission of the United Nations.

---

ment also has an initiative to maintain a common visual theme on its websites, and the 2006 census website observed these basic standards (Laroche, 2005). The Government On-Line effort also included study of security and encryption protocols—an infrastructure on which Statistics Canada was able to piggyback. Similarly, the Australian Electronic Transaction Act of 1999 required agencies to permit electronic communications between citizens and the government (Trewin, 2006). In New Zealand, the "e-government strategy" adopted the goal of making the Internet "the dominant means of enabling ready access to government information and services" by mid-2004 (Smith, 2006).

### Australia

In 2006 (as in previous years), the Australian quinquennial census was conducted on a drop-off–pick-up basis: enumerators delivered forms on the designated Census Night and returned within the next three weeks to pick them up. (Respondents were urged to complete the questionnaire on Census Night, as Australia uses a de facto residence concept.) The questionnaire package delivered to households also included a Census Form Number on the printed questionnaire and a 12-digit eCensus Number in a sealed envelope. Both numbers were needed to use the eCensus application on the Internet. The Australian Bureau of Statistics contracted with IBM to develop its eCensus web application and support systems.

Because of the drop-off–pick-up strategy used for the Australian census, designers needed to provide a mechanism for advising field enumerators that questionnaires in their districts had already been returned online, so that they did not need to do a follow-up visit. Ultimately, the Australian Bureau of Statistics (ABS) settled on notification by text message to enumerator cell phones;[4] this messaging system was part of a larger communications scheme connecting census field staff, central coordinators, and members of the public (who called with inquiries).

Williams (2006) observes that "the 2006 eCensus system was opened to the public just after 8pm on 27 July, with enumerators due to commence delivery of forms on 28 July. The first eCensus respondent submitted their online form at 20:29 on 27 July." In total, ABS experienced a estimated 9 percent response rate via the Internet, representing 775,856 household forms; this slightly exceeded the system's performance in dress rehearsal, in which 7.9 percent of dwellings responded via the Internet. Due to the de facto nature of the census and the encouragement to complete the questionnaire upon receipt, 40.4 percent of all responses received by the Internet came in between 6pm and midnight of the designated Census Night.

Prior to use in 2006, the Internet response option was tested in field tests in 2003 and 2004 and in the 2005 dress rehearsal. Based on the preliminary testing, ABS anticipated—and built its systems to accommodate—a surge of entries on Census Night. Contingency plans, including temporary service interruptions on the eCensus site and public relations messages, were also developed. As it turned out, "the capacity of the system was never really put to the test—with peak load on census night reaching only 15 percent of capacity" (Williams, 2006). ABS also developed contingency plans for malicious denial of service attacks on the census site—deliberate attempts to flood the system in order to shut it down. Mechanisms for monitoring the Internet service providers of incoming log-in attempts were put in place and, "in cases where these attacks could not have been dealt with quickly, public relations messages would have firstly assured the public that their census information is secure and secondly provide information about alternatives

---

[4]A text message was also generated and sent to enumerators if a questionnaire was received by mail and processed.

such as delaying use of the eCensus system or using the paper census form." However, no such denial of service attack was detected.

It is useful to note that Australia is effectively a long-form-only census—using only one questionnaire—rather than a distinction between short- and long-form samples or the 2010 U.S. census short-form-only model.

## Canada

The 2006 Canadian census was the first to offer an online response option.[5] Every paper questionnaire sent by mail or dropped off by enumerators bore a 15-digit Internet Access Code (five groups of three digits) at the upper right of the questionnaire. A banner instruction immediately before "Step A" of the questionnaire read "COMPLETE YOUR FORM ON-LINE OR ON PAPER," and the first question advised respondents that they could complete the form online at a website (http://www.census2006.ca) using the Internet Access Code printed on the form.[6] A follow-up instruction to that option reminded online respondents, "Do not mail back your paper questionnaire."

Online response was permitted for both the census short-form (8 questions) and long-form (53 questions) instruments. The online questionnaire could be rendered in either English or French, and the two languages could be toggled back and forth during the course of completing the online form. The Internet form was designed so that "no software trace (footprint) was left on [a respondent's] computer" once they had submitted it online. However, persons replying to the Canadian long-form questionnaire could indicate that they wished to pause and resume the questionnaire later; they were prompted to create a password and—upon logging back onto the census site—could resume the questionnaire where they left off. If they did not resume the form within some set period of time, though, the partial form was submitted for processing (Statistics Canada, 2007).

Prior to Statistics Canada's designated cutoff date to begin nonresponse follow-up activities, 22 percent of returned questionnaires had been returned online; overall, by the end of August 2006, the online response rate

---

[5]Dolson (2006) describes the multiple response modes offered in the 2006 Canadian census: "Respondents had a choice to respond [to a paper questionnaire sent by mail or dropped off by an enumerator] by either Internet or mail. Some data were collected by personal or CATI interviews. As well, respondents to the long[-form] questionnaire could either reply to the income questions or give Statistics Canada permission to link to their tax records to obtain these data."

[6]Respondents who lost the paper form could call a Census help line to request a new paper questionnaire or an Internet Access Code; alternatively, help line operators could also administer the questionnaire during the phone call. Responses generated though the help line—whether paper, Internet, or direct interview—incurred an extra processing step: matching against an address register to determine the link to a geographic location (Dolson, 2006).

stood at 18.5 percent. Large households (5 or more people) were more likely to invoke the online option (26 percent) than smaller households, including single-member households (of which only 13.5 percent returned the form online). Online response rates did not seem to vary by form type (short or long form), but did vary by province: Alberta experienced the highest online response rate (21.4 percent) and the Northwest Territories and Nunavut—both of which are principally enumerated by personal visit rather than mail—the lowest (13.6 and 0.0 percent, respectively).

The 18.5 percent overall online response rate was consistent with expectations developed based on a 2004 census test using an Internet response option in parts of four provinces, as well as an Internet response experiment conducted as part of the 2001 census. Based on these pretests, Statistics Canada anticipated a 20 percent Internet share in 2006. Significantly, the 2004 test also led Statistics Canada to expect—and plan for—temporal patterns in questionnaire response. Like the U.S. census, Canadian census forms are delivered a few weeks before a designated reference date (Census Day); in the case of the 2006 Canadian census, Census Day was May 16. Based on the testing, Statistics Canada anticipated an early peak in online returns upon the first mailout in early March, with declining amounts until Census Day itself, at which point heightened publicity could be expected to create another response spike. Consistent with expectations, about 15 percent of responses received via the Internet came in on May 16 itself; system managers were able to devise a "graceful deferral" system on Census Day itself to limit the load on census servers.

In terms of data quality, Statistics Canada determined that Internet questionnaires produced much lower item nonresponse rates than did paper questionnaire responses: item nonresponse for paper questionnaires was 102 times higher than Internet questionnaires for short-form responses and 10 times higher for long-form responses. It was also determined that the Internet responses had lower failure rates during basic data editing than the paper forms (Duquet and Gilmour, 2007). In part, this may be due to the use of data confirmation steps that are not possible on a paper form. The Internet short-form questionnaire (as well as computer-assisted forms used in nonresponse follow-up) prompted respondents to confirm the age of household members based on what had already been entered as their dates of birth (rather than answer both questions separately and potentially have a mismatch). The section of the Internet long-form questionnaire on household income also compiled the answers that had already been collected and presented them to the respondents for review and—if necessary—correction.[7] Use of the Internet option may also have saved costs in nonresponse follow-

---

[7]These editing steps are described in Statistics Canada summary of changes in the 2006 census, available at http://www.statcan.ca/english/sdds/document/3901_D17_T9_V1_E.pdf.

up due to the inherent limitation of space on the paper form: the version of the Internet instrument tested in 2004 permitted listings of up to 36 people, compared with the paper form's limit of information for 6 household members and names only for an additional 4 persons (Laroche, 2005).

During the conduct of the 2006 census, Statistics Canada also performed an experiment on targeting the Internet response option to particularly receptive audiences. This study—somewhat similar to the U.S. census tests in 2003 and 2005—was intended to suggest whether households "in geographic areas with a very high Internet penetration rate" might best—and less expensively—be contacted with only a letter and an Internet Access Code (but no questionnaire). As summarized by Statistics Canada (2007:12):

> A model was developed to identify *a priori* areas that include a significant number of dwellings likely to answer the Census online. Households in this study, called the Push Strategy, received only a letter instead of a paper questionnaire. These households were asked to complete their questionnaire online. The letter also included a 1-800 telephone number, which respondents could call for information about the study or to request a paper questionnaire. A preliminary sample of 40,000 households in mail-out areas was selected for this study. This sample was split randomly into two groups of 20,000 households each in order to create a control group [which received a paper questionnaire]. . . . The method was quite effective since the Internet response rate of the Push sample was 2.6 times more than the control group and 3.4 times more than the general population.

The Internet questionnaire used in the 2004 Canadian census test differed significantly from its paper counterpart in its approach to obtaining the basic resident count at a household. The paper questionnaire presents respondents with a set of detailed instructions of who should and should not be included in a household count and then asks for a roster of names. However, the Internet version asked respondents to complete a roster first and then used three follow-up questions—based on the instructions from the paper form—to guide respondents through the process of excluding temporary residents or foreign visitors from the final roster (Larouche, 2005). Whether this feature was also implemented in the final 2006 census Internet instrument is unclear.

Deemed a success in 2006, the online response option is slated for use in the 2011 Canadian census, with the hope of boosting online response to as much as 40 percent. Though definitely not a set policy, Duquet and Gilmour (2007) suggest Statistics Canada's eventual vision for Internet collection in the census, in which an invitation to complete the census online (presumably with an Internet Access Code or the like) and in which a paper questionnaire is mailed only if the household specifically requests one or fails to respond to the initial invitation. Toward that end, Statistics Canada (2007) suggests that it may use its Push Strategy—tested in 2006—on a somewhat larger basis in 2011.

## Japan

Alone in these examples—save for the U.S. 2010 census—Japan elected not to allow online response in its 2005 quinquennial census. For 2005, Kurihara (2004) reports that the Japanese Statistics Bureau sought to improve the information technology infrastructure of the census by rebuilding its internal geographic information system, testing the use of optical character recognition (OCR) of handwritten responses, and redesigning the user interface to obtain and work with small-area census data.[8]

## New Zealand

Like the Australian census, the New Zealand quinquennial census is collected primarily by enumerators dropping off questionnaires and returning at a future date to collect them. Since 1996, New Zealand census questionnaires have been made available in an English-only or bilingual (English/Maori) version, the latter of which uses a "swim-lane" design that is a model for the bilingual English/Spanish form the Census Bureau plans to use in some areas in 2010. For 2006, to better meet perceived user needs, Statistics New Zealand planned an Internet response. However, it purposely did so without "attempt[ing] to leverage efficiency gains in any of the traditional census processes" or forecasting a desired Internet response rate target: plans were made to complete the census using traditional methods, and such responses by the Internet as were completed were deemed "a longer-term investment in improving participation" in later censuses (Smith, 2006). Furthermore, "it was recognized that there would not be financial savings in its implementation in the 2006 Census" (Statistics New Zealand, 2007).

In implementing the Internet response option, Statistics New Zealand (2007) decided not to aggressively promote the option. Instead, the agency chose to rely on limited promotion "through selected high-usage Internet sites only" and—principally—on advocacy from the enumerators assigned to drop off the census forms. As part of their training, census enumerators were

---

[8] On the second of these points—optical character recognition—it is worthwhile to note that this was a major test built into the conduct of the census itself. The specific objective was to determine whether completely automated OCR was sufficiently reliable or whether clerical checks of each questionnaire were still needed. One question—the destination of one's commute to work—was chosen for the automated testing since the seemingly "free" responses to this category were actually limited to the names of about 3,000 municipalities, making quality comparisons easier. Ultimately, it was concluded that "the accuracy of recognition was not sufficiently high" and that research on fully automated recognition would have to continue (Kurihara, 2004:4).

allowed to go through the online response questionnaire themselves; this was deliberately done so that they would be familiar with the requirements and so could accurately inform people in their household workload of the capability to complete the form online. When they visited the households to drop off the questionnaire, they also offered an envelope containing an ePIN identification number in order to use the Internet response option.

The online questionnaire allowed respondents to use either English or Maori. As with the enumerator-dropoff-and-return Australian census, mobile phone text messages were sent to individual enumerators after Internet responses were received, so that those households could be removed from the enumerator's visit workload.

Statistics New Zealand (2007) concluded that "despite very low promotion . . . the online option was very successful, not only in terms of the uptake" (7 percent of responses, or about 400,000 forms, via the Internet) "but an almost completely trouble-free operation." The agency plans to use the Internet response option again—with more active promotion—in 2011.

Prior to implementing the online response option in 2006, the Internet option was included in field tests in March and November 2003 as well as the 2005 dress rehearsal.

## Singapore

In 2000, Singapore transitioned from a traditional census model to a register-based approach. The Household Registration Database (HRD) was developed in 1996 from administrative records as well as 1990 census returns. Hence, the 2000 Singapore census became a sample survey, intended to cover 20 percent of the population, to ask for information not included in the basic register data. These data items included relationship between members of a household, religion, and transportation/commute mode. To carry out this smaller scale survey, the Singapore Department of Statistics adopted a multimode approach. Sample households were invited to complete the form online; if they did not do so by a particular cutoff date, then computer-assisted telephone interviewing (CATI) was attempted. Barring that, trained enumerators were sent out to conduct face-to-face interviews with households that were not reached by either electronic means.

As summarized in a discussion paper for a 2003 census conference,[9] the online response option required respondents to log in using a user id and password, presumably provided in a mailing or through other contact. Once logged on, "basic data already available in the pre-Census database would be displayed" and "the respondent would then proceed to fill up the rest of the questionnaire on-line." Provision was made for respondents to pause

---

[9] http://www.ancsdaap.org/cencon2003/Papers/Singapore/Singapore.pdf.

the interview, save their results, and return at a later time to complete the questions. "Simple on-line checks were included and respondents would be prompted to re-enter the data if the information is incorrect or inconsistent."

Ultimately, about 15 percent of the households in the sample completed the 2000 census form online,[10] and the multimode approach was considered a success.

## Spain

The 2001 Spanish decennial census incorporated two main technological developments in the area of response methodology. One was preprinting of some questionnaire items—including name, sex, birth date, and place of birth—based on entries in Padrón, the Spanish Population Register. Hence, for these questions, respondents confirmed or updated the entries rather than working from purely blank spaces. The second was an Internet response option.

The two technical changes interacted in defining the way respondents were authenticated in order to use the online questionnaire. Those users with no changes to make in the pre-printed Padrón data could enter two personalized "keys" included in the mailing with the census form; alternately, they could access the form if their web browser was equipped with a certain "electronic certificate"—essentially, a digital signature obtained through another agency of the government. Users who wished to update the Padrón information had to have this type of electronic certificate in order to use the Internet form (Moraleda, 2006).

The need for an electronic certificate played some role in dampening the response rate via the Internet. Only 1 percent of households (13,818) completed the form online, of which 29.9 percent authenticated using the certificate. More than this number of households—16,238—attempted to use the Internet census questionnaire to update their Padrón information but gave up because they lacked the requisite certificate (Moraleda, 2006).

The Internet questionnaire application was designed to accommodate completion of the form at multiple sittings: partial information could be saved and then revisited later before submitting a finished questionnaire. The Spanish Internet response option was also available in Spain's co-official languages as well as English, French, German, and Arabic.

## Switzerland

Along with Spain, Switzerland was the other European census to permit online responses as part of its e-Census initiative for the first time in 2001.

---

[10] http://www.singstat.gov.sg/pubn/popn/c2000sr4/coverage.pdf.

Buscher and Stamm (2001:1–2) credited the creation of a government "Service Centre" for managing information technology as a final impetus for allowing online responses—a decision made even though Swiss census officials knew that "only a minority of the Swiss population currently have Internet access." The Swiss Federal Statistical Office reasoned that "electronic communication options are increasingly expected by potential users" and that the "PR and advertising impact of an Internet solution would be highly beneficial for the Census." As in the New Zealand experience, the move was also made with gaining experience with new technology as the guiding goal: "the purpose was to see how far using the Internet could boost the efficiency of data entry and data quality while possibly cutting costs."

Because the Swiss e-Census relied on Service Centre networks, eligibility to file under the e-Census was limited to those communes or regions that had already opted to use the Service Centre equipment; this represented about 90 percent of the total population. Online questionnaires could be administered in German, French, or Italian.

The Swiss online response form was launched on November 27, 2000, and was operated until March 25, 2001; Census Day in the 2000 Swiss census was December 5, 2000. Buscher and Stamm (2001:5) report that "apart from two minor down-times during the first few days of operations, due to high visitor numbers and a server configuration which had not yet been optimized, the e-census ran smoothly, with no security problems throughout the four-month operating period." In all, 281,000 questionnaires (4.2 percent return rate) were completed via the Internet—just under 90 percent of those received during the first three weeks of operation. However, Swiss census officials also found that the form had a curiosity factor: about 20 percent of hits on the questionnaire site seemed to be "tourists" who "wanted to have a quick look at the e-census without attempting to enter their data." Demographically, Internet responses from younger middle-class men were more likely than from other groups but not so much so as to suggests "a major 'digital divide' in Swiss society" (Buscher and Stamm, 2001:7). About 10 percent of visitors to the site were unable to successfully log in to fill out the data: Buscher and Stamm (2001:6) do not describe the log-in procedure, noting only that "while it guaranteed maximum security, was also fairly complicated."

## United Kingdom

The initial design document for the 2011 decennial census of England and Wales (Office for National Statistics, 2004) signaled the intent to use an online response option. Adding the Internet option is considered a useful step in improving the overall response rate, but the Office for National

Statistics (2004:10) recognizes that the option will not immediately cut the cost of the census:

> By increasing the take-up of Internet completion, real cost and time savings could be made by reducing the quantity of paper forms to be captured and processed. Although we would seek to maximize the Internet response in order to realize the potential savings there is no guarantee of success, particularly since among the hard-to-count populations (such as the elderly) there would be significantly lower levels of take-up.

The Office for National Statistics conducted its first major pre-2011 field test in May 2007 with a sample of about 100,000 households. A major focus of the test was to evaluate new residence and national identity questions. However, the 2007 test did not include an Internet response option. A "frequently asked questions" list for the 2007 test posted on the Office for National Statistics website explained that, "as this is a Census Test, resources are limited especially for the large expense to provide a facility to complete the questionnaire online." Nonetheless, the user was reassured that "it is proposed that a facility to complete the questionnaire online will be available for the Census in 2011."[11]

---

[11] http://www.statistics.gov.uk/census/2011Census/2011Project/pdfs/2007TestFAQsEnglish.pdf.

Part III

# Letter Report (February 19, 2009)

# Letter Report

*Addressed to Thomas L. Mesenbourg, acting director and deputy director,*
*U.S. Census Bureau*

This letter relates to plans for tests and experiments planned for the 2010 census. We write to call your attention to several time-sensitive concerns: (1) three crucial topics that should be included in the experimentation during the 2010 census, (2) testing plans preliminary to the census; (3) the retention of 2010 census data, and (4) the designs of the experiments currently planned for 2010.

## BACKGROUND

The Panel on the Design of the 2010 Census Program of Evaluations and Experiments (CPEX) has a broad charge:

> . . . [to] consider priorities for evaluation and experimentation in the 2010 census. [The panel] will also consider the design and documentation of the Master Address File and operational databases to facilitate research and evaluation, the design of experiments to embed in the 2010 census, the design of evaluations of the 2010 census processes, and what can be learned from the pre-2010 testing that was conducted in 2003–2006 to enhance the testing to be conducted in 2012–2016 to support census planning for 2020. Topic areas for research, evaluation, and testing that would come within the panel's scope include questionnaire design, address updating, nonresponse follow-up, coverage follow-up, unduplication of housing units and residents, editing and imputation procedures, and other census operations. Evaluations of data quality would also be within scope. . .

Pursuant to this charge, the panel transmitted an interim report providing general priorities for the CPEX program to the Census Bureau in late

2007 (National Research Council, 2008b) and plans to issue a final report in fall 2009.

The panel met most recently on November 10–11, 2008. At that meeting, Census Bureau staff briefed the panel about the topics that it had chosen for inclusion in the 2010 CPEX program and presented the outlines of the designs for the experiments to be included in the 2010 census. On the basis of those briefings and subsequent discussion, and given the relatively late timing of our final report in the census experimentation planning cycle, the purpose of this letter is to continue to fulfill our charge by providing timely analysis and recommendations for the CPEX program.

## EXPERIMENTATION DURING THE 2010 CENSUS: MISSING TOPICS

A key objective of our interim report (National Research Council, 2008b) was to suggest priority topics for experimentation during the census. In particular, we urged that the topics chosen for experimentation have a direct bearing on visions for the 2020 census (however preliminary) so that they can serve as a first step for research in the intercensal period. We also explicitly recommended that the 2010 experiments be chosen to examine issues with the potential to achieve substantial cost reductions or important improvements in data quality in 2020.

In November 2008, the panel was informed that the Census Bureau has chosen topics for four experiments to be conducted during the 2010 decennial census: (1) a nonresponse follow-up contact strategy experiment, (2) a privacy notification experiment, (3) an alternative questionnaire experiment, and (4) a deadline messaging and compressed schedule experiment. We are deeply concerned that although the topics selected by the Bureau are of interest, they are not grounded in a vision for 2020, nor are they directly linked to cost or data quality concerns. At the same time, we are concerned that two topics with strong potential effects on cost and quality and overall importance for 2020 that we discussed in our interim report are absent from the Bureau's experimentation plans: Internet data collection and the use of administrative records. We reemphasize that these two areas of research are critically important. In addition, we believe that a very different alternative questionnaire experiment—one that tries multiple approaches to improve collection of census residence information—would be invaluable for the future of census questionnaire design.

### Internet Experimentation

The use of the Internet for data collection in the decennial census presents important opportunities for cost reductions and improvements in data quality. These include cost savings through the reduction in the number of forms that have to be scanned or keyed for data entry, reduction in the processing of requests for mailing of foreign language questionnaires, and savings in field work as a result of more prompt receipt of individual data. Use of the Internet may also yield quality improvements through easier access to foreign language questionnaires and online editing of census responses. Importantly, the use of online response would avoid the social cost of the Census Bureau's appearing to be out of step with modern data collection and computing environments.

An experiment in the 2010 census would provide a unique opportunity for examining the use of the Internet for decennial census data collection. A key issue that needs to be explored in an experiment is how large a fraction of the population can be induced in a census environment to use the Internet as a response option, while not at the same time greatly increasing the possibility of disclosure or incurring other security problems. Therefore, we strongly recommend a 2010 census Internet response experiment to help determine ways to increase the likelihood of Internet response in 2020 and possibly also learn how to minimize any associated negative effects. This test should include a "push Internet" option as one of the experimental treatments whereby the initial mail contact strongly encourages Internet response, perhaps even by excluding a paper questionnaire from that initial mailing. Such an experiment could also address the quality of the data collected through the Internet, including for those requiring foreign language questionnaires for whom the Internet may provide a convenient multi-language option.

We recognize that the basic steps to implement an Internet experiment in 2010 are nontrivial: the design and testing of an online version of the census questionnaire, the development of protocols that protect census respondents from disclosure of information, and the integration of online returns with other census operations. However, the panel is confident that the challenges can be overcome, even within a tight time frame, as they were when the Census Bureau added a limited online response option in 2000. In addition to the Census Bureau's own experience with Internet questionnaire development in the 2000 census, the experience of other countries in developing security protocols for online census response (including the 2006 Canadian census) can be tapped as the Census Bureau develops privacy safeguards for online response in planning such an experiment.

## Use of Administrative Records

Administrative records offer substantial potential for both census cost reduction and quality improvements. Administrative records could be used to dramatically reduce the cost of nonresponse follow-up and improve the quality of the resulting data collected by avoiding inaccuracies in "last re-sort" enumerations (often supplied by proxy respondents, such as neighbors or landlords) and by providing higher quality information than is currently supplied by whole-person and whole-household imputation. (An admittedly radical eventual possibility for the use of administrative records would be avoidance of nonresponse follow-up altogether for a large percentage of U.S. households.)

In addition, administrative records could be used to target the imple-mentation of census processes. A key example is that administrative records could identify areas in which the Master Address File (MAF) is deficient, by basing that determination on the difference between the address counts from a merged list of addresses from administrative records and the counts from the MAF, and therefore in need of an address canvass check prior to the decennial census. This approach could dramatically reduce the costs of the currently 100 percent application of the address canvassing operation. One could also use the discrepancy between a household count from the cen-sus and that from administrative records to prioritize the implementation of coverage follow-up interviews. Finally, administrative records could be used to assist in reducing the field work in following up nonmatching cases of the P-sample in coverage measurement.

Although wide-scale use of administrative records to substitute for non-response follow-up would almost certainly require a change in legislation, the potential benefits of increased use of records in census processes should be studied in order to estimate the extent to which such changes would be economically and statistically desirable. Given that the use of administrative records in such a manner provides one of the few opportunities to substan-tially reduce census field costs in 2020, it deserves serious attention in the planned 2010 experiments.

It is important to note that most of the above possibilities for research on administrative records might be properly considered priorities for "evalua-tion" rather than "experimentation" since they would not require additional or special field data collection. (They would, however, require the careful retention of household-level census process data, such as we recommend be-low.) Yet although a great deal about the utility of administrative records can be learned from post hoc study of data retained during the census, there are potentially useful possibilities for limited, experimental field work in 2010. For instance, with regard to the use of administrative records as a substi-tute for late-stage field enumeration, one possible experiment would involve variations in nonresponse follow-up or coverage follow-up protocols under which the number or format of follow-up interviews depended on admin-istrative records information (either on an individual household basis or on an area basis). Such an experiment would involve a significant expansion of the nonresponse follow-up contact strategy experiment (discussed below).

Though "administrative records" in the census context are generally thought to be national-level constructs—drawing information from, for example, Social Security Administration registers—a complete evaluation of records-based methods should also assess the quality of the records maintained by "group quarters" facilities, such as prisons, health care facilities, and college residence halls. Because these facility records were used by census enumerators to count about half of the group quarters' population in the 2000 census, the National Research Council (2006:Table 7-1, pp. 238–240) suggested that the Census Bureau "undertake a continuing research effort to assess the accessibility of facility records at group quarters facilities and to determine whether the existing data systems meet census data collection needs." We endorse this suggestion as it is an essential step to assessing the possibilities for using administrative records to supplement or, as necessary, replace traditional enumeration in group quarters. Assessing the alternative or "home" address information available from facility records is also critical to addressing such long-standing questions as the degree to which college students are counted at both their schools and their parental homes and whether it is feasible to define a "home address" for persons under correctional supervision.

## Census Residence

The 2010 census provides a uniquely valuable setting for a comprehensive experiment involving alternative approaches to the current residence rules. The Census Bureau's proposed alternative questionnaire experiment for 2010 does include one treatment group for gathering a limited amount of information on residence (see below). However, given that unclear residence rules and interpretations were likely a major source of census coverage error (both omission and duplication) in the 2000 census (National Research Council, 2004a), the Panel on Residence Rules in the Decennial Census (National Research Council, 2006) suggested various alternative approaches to collecting information on census residence. In particular, that panel's report proposed a major change from the Census Bureau's traditional approach of relying on a dense set of instructions at the start of the census form to one of asking a set of guided questions that breaks the large cognitive task of deciding one's household composition into smaller pieces. At that panel's urging, the Census Bureau tested a preliminary version of a "worksheet" approach to the residence question in 2005, yet no further work on residence is planned in 2010.

The single treatment group in the proposed alternative questionnaire experiment—anchored to one of the coverage probe questions—falls short of the general "any residence elsewhere" query that the National Research Council (2006) recommended be asked of the general population in a 2010 census experiment and asked of all group quarters (e.g., medical facilities and college housing) residents in the 2010 census itself. The current plans for this limited experiment also do not appear to include the follow-up activities needed to make best use of whatever information might be gained. The proposed single treatment group also falls short of the 2006 report's suggestion to experiment with a de facto or "current residence" question—and add a corresponding de jure or "usual residence" question to the American Community Survey—so that differences in estimates between the two programs due to their differing residence standards could be assessed. Innovative (and more accurate) handling of residence concepts is clearly a research question for which several alternatives need to be tested, and subsequently refined and retested, in order to achieve substantial gains over the Bureau's current approaches.

These three research areas—Internet data collection, the use of administrative records, and questionnaire redesign for residence rules—are ones for which important benefits could be obtained through increases in census data quality or decreases in census costs or both. In the panel's assessment, the 2010 CPEX program should include work on these topics in order to ensure early progress in the 2020 census testing cycle. Therefore, we strongly urge that these topics be included as subjects for experiments in conjunction with the 2010 census.

## SYSTEMS TESTING AND SIMULATION PRIOR TO THE 2010 CENSUS

The panel is concerned that the Census Bureau's operational test plans for the 2010 census are insufficient. We are particularly concerned with the Bureau's capacity to identify potential failure modes in the field data collection components of the 2010 census process. We appreciate that the Census Bureau has had to substantially revise its plans for decennial census nonresponse follow-up. Initial plans to use handheld computers for nonresponse follow-up and to have the operational control system for field data collection developed by a contractor have been dropped in favor of a return to a paper-based nonresponse follow-up operation and a return to an operational control system for field data collection that will be developed in house (presumably by revising the system developed for the 2000 census).

Given the complexity of conducting the decennial census, it has long been deemed essential to have a complete test "dress rehearsal" two years prior to the census so that flaws can be detected and corrected. Given the need to redesign the field data collection plan at this late stage, the census dress rehearsal conducted in 2008 was essentially limited to a test of the mailout/mailback portion of the census process, with no testing of the nonresponse follow-up, coverage follow-up operations, or many other component processes.

The Census Bureau acknowledges that the dress rehearsal provided an inadequate test of the 2010 census processes. As a remedy, it has scheduled a number of small field tests of various components and sub-systems of the census process chain to attempt to identify as many potential flaws as possible prior to implementation. However, given that the operational control system for the field data collection system will not be ready until the summer or fall of 2009, the Census Bureau has decided against a comprehensive test of the entire field data collection process due to the lack of time to design and carry out such a test.

The panel believes that this testing strategy puts the Census Bureau in an extremely risky position should there be flaws in the census process that involve interactions of the many components and subsystems. Testing the interfaces between individual components of a system (e.g., $A \rightarrow B$, $B \rightarrow C$, $C \rightarrow D$) can produce useful information and detect unseen problems. But the Bureau's testing plan creates risks by not adequately testing subsystems (e.g., $A \rightarrow B \rightarrow C$) or complete systems. Errors at this level may not be evident in any single component test but could result in major delays and impair data quality.

Concern over the lack of time or resources to conduct a more comprehensive test is understandable, but it does not override the compelling argument for carrying out such a test. The Census Bureau needs to perform as full and realistic an operational test of all nonresponse follow-up systems as possible. The consequences of failure to identify substantial problems in the interfaces between system components could be dire, ranging from moderate to severe impacts on the quality, costs, and timeliness of census counts for important purposes like redistricting and allocation of funds.

The panel strongly recommends that the Census Bureau try to fit into its schedule a comprehensive test of the entire operational control system for field data collection as soon as feasible after plans for this system become available. We recognize the enormous constraints in planning and accomplishing such testing. Because of these constraints, it may well be necessary in the overall testing to simulate portions of the process based on the specifications for information flows at the interface between component parts of the process. If such simulation is judged to be necessary, then additional field testing of the simulated components of nonresponse follow-up should be carried out.

Ideally, tests should be conducted in enough time to detect—and correct—any problems. But if time is too short to allow for a full cycle of test and correction, earlier detection of defects or inefficiencies can still be vital. Even if a flaw is discovered too late to be addressed in a pre-tested, systematic way, some contingency planning will likely be able to greatly reduce any negative consequences for the census itself.

## RETENTION OF DATA

Since 1985 several National Research Council panels on the decennial census have called for the development of a "master trace sample" database. Such a database would retain the crucial elements of the census procedural history for a sample of addresses to support census evaluation studies. A version of a master trace sample was constructed by the Census Bureau following the 2000 census (Hill and Machowski, 2003). This database supported a small number of studies (e.g., Bentley and Tancreto, 2005; Tancreto and Bentley, 2005; West et al., 2005) that began to realize some of the substantial research potential that such a database could provide.

Our panel's interim report recommended that "the Census Bureau should initiate efforts now for planning the general design of a master trace sample database and should plan for retention of the necessary information to support its creation" (National Research Council, 2008b:Rec. 5). To address the efficacy of less common procedures on small subpopulations, a large sample is clearly needed; we also note that given the greatly decreased cost of computer storage and memory, it may now be possible to save and efficiently access the entire procedural history for the entire country. Whatever the sampling rate, it is critical to retain sufficient data, preserving all relevant linkages, so that the result supports the examination of how the decennial census processes functioned for various subpopulations and domains.

As an example, it is important to retain the information as to which addresses on the MAF were added or deleted by which census address improvement operations. Furthermore, given that many fields of the various system files are overwritten continuously during the census, this means that these data archives should retain snapshots of files that will change during the course of census operations, and this should be provided for as frequently as needed. This data archival effort needs to include all parts of the census process, including address list development, nonresponse follow-up, coverage follow-up, group quarters enumeration, data capture and data treatment, and coverage measurement. In addition, it is vital that the schema used in retaining these data be carefully documented so that it is known precisely what is saved in each data field.

Given the rushed development of the operational control system for field data collection, we are especially concerned that provisions be made for retaining data relating to that part of the census. We do not believe that providing for this additional functionality in the operational control system for the field data collection will add appreciably to the current challenge of developing such a system in time for the 2010 census. Furthermore, by guaranteeing access to this information, the Census Bureau would ensure that it could carry out evaluations that would guide the Bureau towards a more effective and cost-efficient design for the 2020 census. Therefore, we recommend that—as systems for the 2010 census are finalized by the Census Bureau and its contractors—appropriate archival outlets be created for all systems, including components of the field data operational control system, so that the relevant data to construct a master trace database or "audit trail" of census processes are retained. Experts in automated audit processes could provide assistance to the Census Bureau in implementing a master trace system.

## DESIGNS FOR CURRENTLY PLANNED EXPERIMENTS

Although we recommend the addition of three topics for experimentation, the Census Bureau's chosen topics for 2010 experiments do concern issues that may be worth pursuing in addition to our recommended ones. However, three of the four census experiments, as currently outlined, suffer from important defects that will limit their effectiveness. Moreover, the Bureau has not carried out explicit studies of the statistical power of these experiments given their proposed designs. We recognize that the clustering inherent in some of the experimental designs complicates the development of such estimates, but it is also the reason that careful estimates of power are necessary. For each experiment, the Census Bureau needs to undertake a study of the statistical power of the design against reasonable alternatives based on anticipated effect sizes. This should be done not only for national-level comparisons, but also for any relevant subgroup comparisons.

Some of the experiments also do not seem to give appropriate attention to "targeting" or oversampling respondents from relevant sociodemographic groups (or geographic areas with large concentrations of such respondents). Not only does lack of targeting reduce the power of those experiments, but it also hinders the ability to learn more about the response by stratifying the analysis by subgroup.

### The Nonresponse Follow-Up Contact Strategy Experiment

The question of interest in this experiment is the impact on census costs and data quality of reducing the number of attempts made in nonresponse follow-up from a maximum of six to either four or five. As currently planned, the experiment will be carried out in three local census offices, comprising about 40,000 housing units. For each office, two treatments and the control will be randomly allocated to crew leader districts, where all enumerators in a district will use the same questionnaire (which provides space for a maximal number of enumeration attempts) but will receive different instructions about how many callbacks to make. To assess the treatments and control, comparisons will be made of the resulting impact on census data quality, measured by the rate of proxy response, the distribution of response outcomes, the item nonresponse rate, and measures of form completeness. The Census Bureau staff have expressed a concern as to whether the findings would be generalizable from the three local census offices, and asked the panel for assistance in selecting local census offices for this experiment. However, our current overriding concern is whether data from only three local office areas can ever be sufficiently generalizable.

In addition to questions about generalizability and statistical power, the panel questions whether the likely reduction in field data collection costs will be sufficient to justify the allocation of resources for an experiment during the 2010 census. The likely impact on census costs might be fairly modest. In the November meeting, the panel suggested that the cost reduction could be estimated on the basis of the frequency of enumerations in 2000 that were successful on the fourth or fifth attempts. The Census Bureau argued that such estimates are misleading due to infrastructure changes that occur during the taking of the census, such as the laying off of enumerators, consolidation of work, and other changes. The panel countered that estimates based on an analysis of 2000 census data, while somewhat flawed due to such changes, would still provide a sense of whether the potential reductions in field costs would be large enough to justify a separate experiment during the 2010 census. Based on such estimates, if the cost reduction seems likely to be, at best, modest, the experiment should be eliminated or redesigned to include assessment of even fewer enumeration attempts or the use of administrative records in lieu of field data collection.

In considering statistical power, 2000 data could have been used to estimate the percentage of housing units that first failed to return their mailed questionnaire, and then were enumerated in the 2000 census on either the fourth or fifth attempt during nonresponse follow-up. In doing so, it may be discovered that the effective sample size for this experiment is too small to provide sufficient power to identify important differences in the above data quality measures (unless such differences are strikingly large). If it is clear that the experiment will not have substantial power to detect reasonable changes to the census data quality measures, and if a two or three-fold increase in the number of local census offices would provide sufficient power, the sample size should be expanded. If no conceivable sample size can provide reasonable statistical power, the experiment would not be useful and should not be done.

One additional argument in favor of an experiment on this topic, if slightly broadened, is that there is an a distinct disadvantage of waiting until six responses are attempted. This disadvantage is that the lag between

Census Day and the day of enumeration increases the number of movers and in general reduces data quality and increases the rate of erroneous enumeration. Assessment of this disadvantage, possibly in conjunction with the coverage measurement program, might be very useful.

### The Privacy Notification Experiment

The privacy notification experiment will assess the effect of a message on the cover letter of the mailing package containing the census questionnaire regarding the uses of census data and the possible use of administrative records. The experiment includes two panels of 10,000 sampled households each (plus a control group without such notification), chosen using strata based on levels of mail response in the 2000 census or in the American Community Survey. The assessment of the three wordings will use response rates, data quality measures, and monitoring of public reaction. The hope is to be able to have reasonable power to identify a difference in overall mailback rate of 1.8 percent. (A one percent reduction in mail response is estimated to cost the Census Bureau $90 million in 2010.)

The panel has three principal concerns with the current design of this experiment. The treatment panels vary only in the wording of one part of the notification message—"Your answers will be used for statistical purposes, and no other purpose" compared with "Your answers will only be used to produce statistics"—raising concerns about how informative the test will actually be regarding individual perceptions of privacy. Second, a longer, second section of the message is identical between the two treatment groups and hints at the possible use of administrative records:

> To improve census results, other government agencies may give us information about your household. The additional information we receive is legally protected under Title 13, like your census answers.

If the objective of the experiment is to assess privacy concerns, it would be beneficial to explore other wordings of this second part of the notification. Instead of a single test of a very limited set of alternative statements in 2010, it would be more useful for the Census Bureau to conduct a series of intercensal tests between 2010 and 2020 that would develop a broad sense of people's sensitivity to privacy concerns and use of administrative records. Such a research program should examine this for sociodemographic subsets of the population.

Another deficiency is that the Census Bureau is not using this opportunity to evaluate the implied tradeoff of the costs incurred from the freedom to use administrative records as a result of the inclusion of such a notification and the benefits from being allowed to do so. That is, while the privacy notification may have the effect of reducing mail response rates, it will at the same time allow for the use of administrative records to reduce costs and improve data quality, for example, by substituting for last-resort and proxy enumeration. Therefore, it seems reasonable to use this opportunity to determine the degree to which administrative records can reduce census costs and improve census data quality and whether such benefits offset the reduction in mail response and the associated increase in the costs of nonresponse follow-up. Possibly, this could be done through the separate administrative records experiment noted above, but bundling this as a single experiment may have some advantages, although it would increase the complexity of the currently planned experiment.

### The Alternative Questionnaire Experiment

There are three parts to the proposed 2010 questionnaire experiment: (a) a comparison of the complete set of questionnaire changes between 2000 and 2010, (b) an attempt to collect an alternative residence address based on answers to a coverage probe question, and (c) alternative formats for the collection of information on race and ethnicity. In part (a), 10,000 housing units will receive a 2000-style census questionnaire. Comparisons will be made to the distribution of responses to the full 2010 census to ascertain what changes between 2000 and 2010 are due to changes in questionnaire format. In part (b), 30,000 housing units will be administrated an alternative questionnaire that will permit respondents to specify a street address if they indicate that the person in question sometimes lives or stays at another location. In part (c), 30,000 housing units in each of 11 panels will be administered various questionnaire formats for the questions on race and ethnicity. Some of these will present slightly different versions of a combined race and Hispanic origin question (the 2010 census questionnaire itself presents them as separate numbered items). Other treatment groups respond to census advisory committee suggestions by permitting multiple and write-in answers to the Hispanic origin question or varying specific examples that are explicitly mentioned in the question (e.g., Taiwanese or Marshallese). It is planned that cognitive testing will be carried out in advance of the experiment to better refine the various alternatives. The forms will be mailed to a random sample of housing units, and initial nonrespondents will receive a replacement questionnaire that mimics the initial questionnaire. The goal of the experiment is not to identify specific alternative formats, but rather to learn more about the general formats that are preferred in order to fold this information into a longer term research program on questionnaire design.

The goals of parts (a) and (b) are not clear to the panel. Consequently, it is hard to judge whether the experimental designs and sample sizes are suitable and whether the experiments are likely to yield useful results. Although the sample size for part (a) may be sufficient to detect any economically important change in overall response rates between these two forms of the questionnaire, it may not be adequate if one wishes to understand how these changes are related to subgroups of the population, size of family, etc. The sample size is also not likely to be adequate if one is attempting to relate specific changes in response patterns to specific living situations, membership in demographic subgroups, etc. Otherwise, interpretation of any changes in response patterns will be limited due to confounding as a result of the several simultaneous changes to the questionnaire. As a result, the benefits for questionnaire design for 2020 will be reduced.

With respect to part (b) we are concerned about adequate power because it was unclear that 30,000 households would provide a large enough number of alternative addresses to be able to determine whether the inclusion of such a question on the census questionnaire would be able to substantially affect the need for the coverage follow-up interview or the accuracy of such an interview if it appeared to be needed. Therefore, some form of targeting— say of areas with a high frequency of seasonal second homes, or of people living in types of group quarters that frequently involve duplication—would be desirable. Second, it was not clear that this part included sufficient provision for gathering follow-up information so as to determine the usefulness of the additional question. That is, although the addition of any question on the census form has an associated cost of processing and a possible decrease in overall data quality, the inclusion of this question could produce higher quality responses as to census residence and/or it could also affect the frequency of coverage follow-up interviews or their accuracy. Therefore, it is important to include plans in the experimental protocol that would attempt to evaluate this tradeoff, since this should be key to making any decisions about the inclusion of such a question in the 2020 census questionnaire.

The race/ethnicity arms of this experiment (part c) involve fine distinctions in question wording that are most applicable to specific demographic subgroups. In particular, a major emphasis in this section is on Hispanic respondents. Therefore, this experiment would greatly benefit from any efforts to target the delivery of the questionnaire to areas with a larger percentage of Hispanic residents. In addition, given the increased use of bilingual questionnaires in the 2010 census to facilitate response for essentially the same population, it would be useful to extend this experiment to examine the impact of such changes on a bilingual version of the census questionnaire.

## The Deadline Messaging and Compressed Schedule Experiment

The key question of this experiment is whether the rate of mail response could be increased as a result of the use of deadline messaging (namely, the use of a notice on the mailing package that the form is required to be returned by a specific date) or a compressed mailing schedule or both. In the experiment, three sampling strata will be used: high, medium, and low mail response areas. Each of the eight study panels will involve 10,000 households. These eight panels are: (1) control, (2) compressed mailing schedule panel, (3–5) three deadline messaging panels, and (6–8) three compressed schedule combined with deadline messaging panels. The three deadline messaging panels have language of varying degrees of sternness related to delays in mailing back the questionnaire. The analysis will focus on response rates, speed of response, and item nonresponse rates. Our only concern about this experiment is the lack of specification of the statistical power.

In summary, as the Census Bureau finalizes its preparations for the 2010 census, the panel believes that the Bureau faces tremendous risk if it does not perform comprehensive systems testing—focused on the interfaces between individual system components and, ideally, involving some field work component. The quality and utility of 2010 census evaluations will also be seriously impaired if census operational systems are not designed to retain procedural data for construction of a master trace database. The Census Bureau has proposed four experiments to be conducted during the 2010 census, but the panel believes that they suffer from design flaws and, significantly, lack connection to potential visions for the 2020 census. The panel suggests that three topics that are given little or no weight in the current CPEX plan—Internet data collection, use of administrative records in various census processes, and elicitation of accurate residence information—have greater potential to decrease the cost and increase the quality of the 2020 census, and so should be built into the 2010 experimental program.

We hope that the information and recommendations in this letter are useful to the Census Bureau. We would be happy to discuss and explain any of these issues at your convenience.

Sincerely,

Lawrence D. Brown, *Chair*
Panel on the Design of the 2010 Census
Program of Evaluations and Experiments (CPEX)

# References

Abramson, F. H. (2004, November 17). *Census 2000 Testing, Experimentation, and Evaluation Program Summary Results.* Washington, DC: U.S. Census Bureau. Available: http://www.census.gov/pred/www/rpts/TXE%20Program%20Summary.pdf.

Anderson, M. J. (1988). *The American Census: A Social History.* New Haven, CT: Yale University Press.

Anderson, M. J. (Ed.) (2000). *Encyclopedia of the U.S. Census.* Washington, DC: CQ Press.

Angueira, T. (2003, May). *Reengineering the Decennial Census: The Baseline Design for 2010.* 2010 Census Planning Memoranda Series, Number 14. Version 1.5. Washington, DC: U.S. Census Bureau.

Bailar, B. A. (2000). Census testing. In M. J. Anderson (Ed.), *Encyclopedia of the U.S. Census,* pp. 62–66. Washington, DC: CQ Press.

Bauder, M. and D. Judson (2003). *Administrative Records Experiment in 2000 (AREX 2000): Household Level Analysis.* Census 2000 Experiment Report. Washington, DC: U.S. Census Bureau.

Bentley, M. and J. G. Tancreto (2005). Census 2000 nonresponse follow-up: Discrepancies in enumerator assigned housing unit status. In *Proceedings of the Section on Survey Research Methods.* Alexandria, VA: American Statistical Association.

Berning, M. A. (2003). *Administrative Records Experiment in 2000 (AREX 2000): Request for Physical Address Evaluation.* Census 2000 Experiment Report. Washington, DC: U.S. Census Bureau.

Berning, M. A. and R. H. Cook (2003). *Administrative Records Experiment in 2000 (AREX 2000): Process Evaluation.* Census 2000 Experiment Report. Washington, DC: U.S. Census Bureau.

Box, G. E. P. and J. S. Hunter (1961). The $2k - p$ fractional factorial designs. part I (Corr. V5 p417). *Technometrics 3*, 311–351.

Brudvig, L. (2003). *Analysis of the Social Security Number Validation Component of the Social Security Number, Privacy Attitudes, and Notification Experiment.* Census 2000 Testing, Experimentation, and Evaluation Program Report. Washington, DC: U.S. Census Bureau.

Buscher, M. and H. Stamm (2001). E-census: The 2000 census on the Internet—concept and stock-taking. Paper contributed to the Second Swiss eGovernment Symposium, August 22, 2001, Zurich.

Bye, B. V. and D. H. Judson (2004). *Results from the Administrative Records Experiment in 2000.* Census 2000 Testing, Experimentation, and Evaluation Program. Topic Report 16. Washington, DC: U.S. Census Bureau.

Cabinet Office (2008, December). *Helping to Shape Tomorrow: The 2011 Census of Population and Housing in England and Wales.* Presented to Parliament by the Minister to the Cabinet Office and laid before the National Assembly for Wales by the Minister for Finance and Public Service Delivery. London: Cabinet Office.

Caspar, R. A. (2003). *Synthesis of Results from the Response Mode and Incentive Experiment.* Census 2000 Testing, Experimentation, and Evaluation Program Report. Washington, DC: U.S. Census Bureau.

Caspar, R. A. (2004). *Results from the Response Mode and Incentive Experiment in 2000.* Census 2000 Testing, Experimentation, and Evaluation Program. Topic Report 18. Washington, DC: U.S. Census Bureau.

Citro, C. F. (2000). Coverage improvement procedures. In M. J. Anderson (Ed.), *Encyclopedia of the U.S. Census*, pp. 101–104. Washington, DC: CQ Press.

Crowley, M. (2003). *Generation X Speaks Out on Civic Engagement and the Decennial Census: An Ethnographic Approach.* Census 2000 Ethnographic Study. Washington, DC: U.S. Census Bureau.

de la Puente, M. and R. McKay (1995). Developing and testing race and ethnic origin questions from the Current Population Survey Supplement on Race and Ethnic Origin. In *Proceedings of the Section on Government Statistics*. Alexandria, VA: American Statistical Association.

Dillman, D. A. (1978). *Mail and Telephone Surveys: The Total Design Method.* New York: Wiley.

Dillman, D. A., J. R. Clark, and M. D. Sinclair (1993a). The 1992 simplified questionnaire test: Effects of questionnaire length, respondent-friendly design and request for Social Security numbers on completion rates. In *Proceedings of the 1993 Bureau of the Census Annual Research Conference*, pp. 8–17. Washington, DC: U.S. Department of Commerce.

Dillman, D. A., J. R. Clark, and M. D. Sinclair (1993b). How prenotice letters, stamped envelopes and reminder postcards affect mailback response rates for census questionnaires. In *Annual Research Conference Proceedings*. Washington, DC: U.S. Department of Commerce.

Dillman, D. A., J. R. Clark, and J. B. Treat (1994). Influence of 13 design factors on completion rates to decennial census questionnaires. Paper presented at the Annual Research Conference of the U.S. Bureau of the Census, Arlington, Va. (March).

Dolson, D. (2006). Efficient multi mode data collection in a census context. In *Proceedings of the Section on Survey Research Methods*. Alexandria, VA: American Statistical Association.

Duquet, L. and G. Gilmour (2007, April). 2006 census online project. Presentation at the Statistics Canada IT Conference.

Durand, E. D. (1910, April 9). The census and the social worker. *The Survey 24*, 81–84.

*Geographical* (2004, October). Counting on the public: Although the 2001 census used new techniques to produce the most complete survey of the UK population, its statisticians still encountered problems [note]. *Geographical*. Reprinted: http://findarticles.com/p/articles/mi_hb3120/is_10_76/ai_n29121905/?tag=content;col1.

Ericksen, E. P., L. F. Estrada, J. W. Tukey, and K. M. Wolter (1991). *Report on the 1990 Decennial Census and the Post-Enumeration Survey*. Washington, DC: U.S. Department of Commerce.

Gerber, E. (2003). *Privacy Schemas and Data Collection: An Ethnographic Account*. Census 2000 Ethnographic Study. Washington, DC: U.S. Census Bureau.

Gerber, E., A. Dajani, and M. A. Scaggs (2002). *An Experiment to Improve Coverage Through Revised Roster Instructions*. Census 2000 Alternative Questionnaire Experiment Report. Washington, DC: U.S. Census Bureau.

Goldfield, E. D. and D. M. Pemberton (2000a). 1950 census. In M. J. Anderson (Ed.), *Encyclopedia of the U.S. Census*, pp. 143–148. Washington, DC: CQ Press.

Goldfield, E. D. and D. M. Pemberton (2000b). 1960 census. In M. J. Anderson (Ed.), *Encyclopedia of the U.S. Census*, pp. 148–153. Washington, DC: CQ Press.

Griffin, D. H., D. P. Fischer, and M. T. Morgan (2001). Testing an Internet response option for the American Community Survey. Paper presented at American Association for Public Opinion Research annual conference, Montreal, May 17–20.

Griffin, D. H. and S. M. Obenski (2001, September 28). *A Demonstration of the Operational Feasibility of the American Community Survey*. Census 2000 Testing, Experimentation, and Evaluation Program Report. Washington, DC: U.S. Census Bureau.

Groves, R. M. (2009, October 7). *2010 Census: A Status Update of Key Decennial Operations*. Prepared statement before the Subcommittee on Federal Financial Management, Government Information, Federal Services and International Security, Committee on Homeland Security and Governmental Affairs, U.S. Senate: U.S. Census Bureau. Accessible: http://www.census.gov/Press-Release/www/releases/pdf/GrovesCensusSenateTestimony10-7.pdf.

Guarino, J. (2001). *Assessing the Impact of Differential Incentives and Alternative Data Collection Modes on Census Response*. Census 2000 Testing, Experimentation, and Evaluation Program Report. Washington, DC: U.S. Census Bureau.

Guarino, J. A., J. M. Hill, and H. F. Woltman (2001). *Analysis of the Social Security Number Notification Component of the Social Security Number, Privacy Attitudes, and Notification Experiment*. Census 2000 Testing, Experimentation, and Evaluation Program Report. Washington, DC: U.S. Census Bureau.

Hacker, J. D. (2000a). 1850 census. In M. J. Anderson (Ed.), *Encyclopedia of the U.S. Census*, pp. 123–125. Washington, DC: CQ Press.

Hacker, J. D. (2000b). 1870 census. In M. J. Anderson (Ed.), *Encyclopedia of the U.S. Census*, pp. 127–129. Washington, DC: CQ Press.

Heimovitz, H. K. (2003). *Administrative Records Experiment in 2000 (AREX 2000): Outcomes Evaluation*. Census 2000 Experiment Report. Washington, DC: U.S. Census Bureau.

Hill, J. M. and J. D. Machowski (2003, September 29). *Master Trace Sample*. Census 2000 Evaluation B.6. Washington, DC: U.S. Census Bureau.

Hill, J. M., J. G. Tancreto, and C. Rothhaas (2006). Experimental treatment results for the age, relationship, and tenure items from the 2005 National Census Test. In *Proceedings of the Section on Survey Research Methods*, pp. 3130–3137. Alexandria, VA: American Statistical Association.

Jackson, A. A. (2008, August 14). *2010 Census Program for Evaluations and Experiments*. 2010 Decennial Census Program Decision Memorandum Series No. 24. Washington, DC: U.S. Census Bureau.

Jenkins, R. (1983). *Procedural History of the 1940 Census of Population and Housing*. Prepared at Center for Demography and Ecology, University of Wisconsin–Madison. Washington, DC: U.S. Census Bureau.

Jenkins, R. M. (2000). 1940 census. In M. J. Anderson (Ed.), *Encyclopedia of the U.S. Census*, pp. 140–143. Washington, DC: CQ Press.

Judson, D. H. and B. Bye (2003, October 21). *Synthesis of Results from the Administrative Records Experiment in 2000 (AREX 2000)*. Washington, DC: U.S. Census Bureau.

Karl, L., E. A. Krejsa, and A. Landreth (2005). Design of the 2004 Coverage Research Followup questionnaire. In *Proceedings of the Section on Survey Research Methods*, pp. 3172–3179. Alexandria, VA: American Statistical Association.

Kim, J., E. T. Huang, and K. Marquis (1998). Evaluation of the 1996 Community Census administrative records file. In *Proceedings of the Section on Survey Research Methods*, pp. 196–201. Alexandria, VA: American Statistical Association.

Kincannon, C. L. (2006, June 6). The 2010 decennial census program. Prepared statement, testimony before the Subcommittee on Federal Financial Management, Government Information, and International Security of the Committee on Homeland Security and Governmental Affairs, U.S. Senate.

Knight, L. M., J. T. Behler, and F. A. Vitrano (2005). Operational assessment of the 2004 Coverage Research Followup. In *Proceedings of the Section on Survey Research Methods*, pp. 3228–3234. Alexandria, VA: American Statistical Association.

Kurihara, N. (2004). Some issues in the 2005 population census of Japan. Statistics Bureau, Japan.

Laroche, D. (2005). *Evaluation report on Internet option of 2004 census test: Characteristics of electronic questionnaires, non-response rates, follow-up rates, and qualitative issues*. Working Paper 23, Work Session on Statistical Data Editing. Ottawa: United Nations Statistical Commission and Economic Commission for Europe and Conference of European Statisticians.

Larwood, L. and S. Trentham (2004). *Results from the Social Security Number, Privacy Attitudes, and Notification Experiment in Census 2000*. Census 2000 Testing, Experimentation, and Evaluation Program. Topic Report 19. Washington, DC: U.S. Census Bureau.

Magnuson, D. L. (2000a). Decennial censuses: 1910 census. In M. J. Anderson (Ed.), *Encyclopedia of the U.S. Census*, pp. 135–136. Washington, DC: CQ Press.

Magnuson, D. L. (2000b). Decennial censuses: 1930 census. In M. J. Anderson (Ed.), *Encyclopedia of the U.S. Census*, pp. 139–140. Washington, DC: CQ Press.

Martin, E. (2002). *Questionnaire Effects on Reporting of Race and Hispanic Origin: Results of a Replication of the 1990 Mail Short Form in Census 2000*. Census 2000 Alternative Questionnaire Experiment Report. Washington, DC: U.S. Census Bureau.

Martin, E. (2007, July 25). *Final Report of an Experiment: Effects of a Revised Instruction, Deadline, and Final Question Series in the Decennial Mail Short Form*. Research Report Series—Survey Methodology #2007-25 and 2010 Census Memoranda Series. Washington, DC: U.S. Census Bureau.

Martin, E., E. Gerber, and C. Redline (2003, August 28). *Synthesis Report: Census 2000 Alternative Questionnaire Experiment*. Census 2000 Alternative Questionnaire Experiment Report. Washington, DC: U.S. Census Bureau.

Martin, E., D. Sheppard, M. Bentley, and C. Bennett (2003). *Results of 2003 National Census Test of Race and Hispanic Questions.* Unpublished report, dated October 1. Washington, DC: U.S. Census Bureau.

McKay, R. and M. de la Puente (1995). Research improves questions. *Civil Rights Journal 1*(1).

McMillen, D. (2000). Apportionment and redistricting. In M. J. Anderson (Ed.), *Encyclopedia of the U.S. Census*, pp. 34–42. Washington, DC: CQ Press.

MITRE Corporation (2007, January 17). An assessment of the risks, costs, and benefits of including the Internet as a response option in the 2010 decennial census. Version 2.0. Report prepared for the U.S. Census Bureau.

Moraleda, A. G. (2006). *Census data collecting by Internet: Spanish experience in 2001.* Working Paper 12, Work Session on Electronic Raw Data Reporting. Geneva: United Nations Statistical Commission and Economic Commission for Europe and Conference of European Statisticians.

Mule, T. (2002, December 31). *A.C.E. Revision II Results: Further Study of Person Duplication.* DSSD Revised A.C.E. Estimates Memorandum Series PP-51. Washington, DC: U.S. Census Bureau.

Mule, T., T. Schellhammer, D. Malec, and J. Maples (2007, March 6). *Using Continuous Variables as Modeling Covariates for Net Coverage Estimation.* DSSD 2010 Census Coverage Measurement Memorandum Series #2010-E-09-RI. Washington, DC: U.S. Census Bureau.

Mulry-Liggan, M. (1986). *Overview and Summary of the Forward Trace Study.* Statistical Research Division Report Series, Number Census/SRD/RR-86-24. Washington, DC: U.S. Census Bureau.

Nash, F. F. (2000). *Overview of the Duplicate Housing Unit Operations.* Washington, DC: U.S. Census Bureau.

National Research Council (1978). *Counting the People in 1980: An Appraisal of Census Plans.* Panel on Decennial Census Plans, Committee on National Statistics. Washington, DC: National Academy Press.

National Research Council (1985). *The Bicentennial Census: New Directions for Methodology in 1990.* Panel on Decennial Census Methodology, Constance F. Citro and Michael L. Cohen, eds., Committee on National Statistics. Washington, DC: National Academy Press.

National Research Council (1988). *Priorities for the 1990 Census: Research, Evaluation, and Experimentation (REX) Program.* Panel on Decennial Census Methodology, Constance F. Citro and Michael L. Cohen, eds., Committee on National Statistics. Washington, DC: National Academy Press.

National Research Council (1994). *Counting People in the Information Age.* Panel to Evaluate Alternative Census Methods, Duane L. Steffey and Norman M. Bradburn, eds., Committee on National Statistics. Washington, DC: National Academy Press.

National Research Council (1995). *Modernizing the U.S. Census.* Panel on Census Requirements in the Year 2000 and Beyond, Barry Edmonston and Charles Schultze, eds., Committee on National Statistics. Washington, DC: National Academy Press.

National Research Council (1997). *Preparing for the 2000 Census: Interim Report II.* Panel to Evaluate Alternative Census Methodologies, Andrew A. White and Keith F. Rust, eds., Committee on National Statistics. Washington, DC: National Academy Press.

National Research Council (1999). *Measuring a Changing Nation: Modern Methods for the 2000 Census.* Panel on Alternative Census Methodologies, Michael L. Cohen, Andrew A. White, and Keith F. Rust, eds., Committee on National Statistics. Washington, DC: National Academy Press.

National Research Council (2003). *Planning the 2010 Census: Second Interim Report.* Panel on Research on Future Census Methods, Daniel L. Cork, Michael L. Cohen, and Benjamin F. King, eds., Committee on National Statistics. Washington, DC: The National Academies Press.

National Research Council (2004a). *The 2000 Census: Counting Under Adversity.* Panel to Review the 2000 Census, Constance F. Citro, Daniel L. Cork, and Janet L. Norwood, eds., Committee on National Statistics. Washington, DC: The National Academies Press.

National Research Council (2004b). *Reengineering the 2010 Census: Risks and Challenges.* Panel on Research on Future Census Methods, Daniel L. Cork, Michael L. Cohen, and Benjamin F. King, eds., Committee on National Statistics. Washington, DC: The National Academies Press.

National Research Council (2006). *Once, Only Once, and in the Right Place: Residence Rules in the Decennial Census.* Panel on Residence Rules in the Decennial Census, Daniel L. Cork and Paul R. Voss, eds., Committee on National Statistics, Division of Behavioral and Social Sciences and Education. Washington, DC: The National Academies Press.

National Research Council (2007). *Using the American Community Survey: Benefits and Challenges.* Panel on the Functionality and Usability of Data from the American Community Survey. Constance F. Citro and Graham Kalton, eds., Committee on National Statistics, Division of Behavioral and Social Sciences and Education. Washington, DC: The National Academies Press.

National Research Council (2008a). *Coverage Measurement in the 2010 Census.* Panel on Correlation Bias and Coverage Measurement in the 2010 Decennial Census, Robert M. Bell and Michael L. Cohen, eds., Committee on National Statistics,

Division of Behavioral and Social Sciences and Education. Washington, DC: The National Academies Press.

National Research Council (2008b). *Experimentation and Evaluation Plans for the 2010 Census.* Panel on the Design of the 2010 Census Program of Evaluations and Experiments, Lawrence D. Brown, Michael L. Cohen, and Daniel L. Cork, eds. Committee on National Statistics, Division of Behavioral and Social Sciences and Education. Washington, DC: The National Academies Press.

National Research Council (2009a, February 19). *Letter from Lawrence D. Brown to Thomas H. Mesenbourg, acting director, U.S. Census Bureau.* Panel on the Design of the 2010 Census Program of Evaluations and Experiments, Lawrence D. Brown, Michael L. Cohen, and Daniel L. Cork, eds. Committee on National Statistics, Division of Behavioral and Social Sciences and Education. Washington, DC: The National Academies Press.

National Research Council (2009b). *Principles and Practices for a Federal Statistical Agency* (4th ed.). Committee on National Statistics, Constance F. Citro, Miron L. Straf, and Margaret E. Martin, eds. Division of Behavioral and Social Sciences and Education. Washington, DC: The National Academies Press.

National Statistical Office, Republic of Korea (2006, November). *E-Census as a New Approach to the Population and Housing Census of Korea.* Paper presented at 11th Meeting of the Heads of National Statistical Offices of East Asian Countries. Tokyo: Korea National Statistical office.

Office for National Statistics (2004, March). *Information Paper: The 2011 Census: A Design for England and Wales.* London: Office for National Statistics.

Pennington, R. A. (2005). Unduplication of persons and housing units in the 2004 Census Test. In *Proceedings of the Section on Survey Research Methods*, pp. 3470–3475. Alexandria, VA: American Statistical Association.

Raglin, D. A. (1998a). The effect of a household-level screening question on the prevalence rate of an item. In *Proceedings of the Section on Survey Research Methods*, pp. 398–403. Alexandria, VA: American Statistical Association.

Raglin, D. A. (1998b). Using the 1996 National Content Survey to measure the impact of alternative skip patterns on the collection of employment data. In *Proceedings of the Section on Survey Research Methods*, pp. 342–407. Alexandria, VA: American Statistical Association.

Redline, C., D. A. Dillman, A. Dajani, and M. A. Scaggs (2002). *The Effects of Altering the Design of Branching Instructions on Navigational Performance in Census 2000.* Census 2000 Alternative Questionnaire Experiment Report. Washington, DC: U.S. Census Bureau.

Reichert, J. W. (2009). Status of the 2010 CPEX evaluations. Presentation to the Panel on the Design of the 2010 Census Program of Experiments and Evaluations, February 17, 2009, Washington, DC.

Rodriguez, C. E. (1994, February 17–18). Challenges and emerging issues: Race and ethnic identity among Latinos. In *Proceedings from the Workshop on Race and Ethnicity Classification: An Assessment of the Federal Standards for Race and Ethnicity Classification*. Washington, DC.

Schneider, S., D. Cantor, P. Segel, C. Arieira, and L. Nguyen (2002). *Response Mode and Incentive Experiment for Census 2000*. Census 2000 Testing, Experimentation, and Evaluation Program Report. Washington, DC: U.S. Census Bureau.

Schwede, L. (2003, August 27). *Complex Households and Relationships in the Decennial Census and in Ethnographic Studies of Six Race/Ethnic Groups*. Census 2000 Testing, Experimentation, and Evaluation Program Report. Washington, DC: U.S. Census Bureau.

Smith, I. (2006). Using both Internet and field collection methods for the 2006 census of population and dwellings. Invited paper for the Population Association of New Zealand Conference, Statistics New Zealand.

Statistics Canada (2007). Census technology: Recent developments and implications on census methodology—census on the net. Paper presented for the tenth session of the Economic Commission for Europe, Conference of European Statisticians, Group of Experts on Population and Housing Censuses. ECE/CES/GE 41/2007/9.

Statistics New Zealand (2007). Implications of the Internet census for the management of field operations. Paper prepared for the Joint UNECE/Eurostat Meeting on Population and Housing Censuses, Kazakhstan, June 4–6, 2007.

Stukel, D. (2008, August). *Projected Census Dates, Funding Requirements and Sources, and Technical Assistance Needs for the 2010 Round of Population and Housing Censuses*. New York: United Nations Statistics Division.

Swain, L., J. D. Drew, B. Lafrance, and K. Lance (1992). The creation of a residential address register for coverage improvement in the 1991 Canadian census. *Survey Methodology 18*, 127–141.

Tancreto, J. G. (2006). An overview of the 2005 National Census Test. In *Proceedings of the Section on Survey Research Methods*, pp. 3764–3771. Alexandria, VA: American Statistical Association.

Tancreto, J. G. and M. Bentley (2005). Determining the effectiveness of multiple nonresponse follow-up contact attempts on response and quality. In *Proceedings of the Section on Survey Research Methods*, pp. 3626–3632. Alexandria, VA: American Statistical Association.

Treat, J. B. (1993). *1993 National Census Test Appeals and Long-Form Experiment, Long-Form Component, Final Report*. Washington, DC: Bureau of the Census, U.S. Department of Commerce.

Treat, J. B., S. Brady, J. A. Bouffard, and C. Stapleton (2003). *2003 National Census Test: The Impact of Alternative Modes and Contact Strategies on Self-Response*. Unpublished report. Washington, DC: U.S. Census Bureau.

Trentham, S. and L. Larwood (2003). *Synthesis of Results from the Social Security Number, Privacy Attitudes, and Notification Experiment.* Census 2000 Testing, Experimentation, and Evaluation Program Report. Washington, DC: U.S. Census Bureau.

Trewin, D. (2006). *How Australia Takes a Census.* Report 2903.0. Canberra: Australian Bureau of Statistics.

United Nations Economic Commission for Europe (2006). *Conference of European Statisticians: Recommendations for the 2010 censuses of population and housing.* Prepared in cooperation with the Statistical Office of the European Communities. ECE/CES/STAT/NONE/2006/4. New York and Geneva: United Nations.

U.S. Census Bureau (1955). *The 1950 Censuses—How They Were Taken.* Procedural Studies of the 1950 Censuses, No. 2; Population, Housing, Agriculture, Irrigation, Drainage. Washington, DC: U.S. Department of Commerce.

U.S. Census Bureau (1963). *Evaluation and Research Program of the U.S. Censuses of Population and Housing, 1960: Background, Procedures, and Forms.* ER 60–1. Washington, DC: U.S. Government Printing Office.

U.S. Census Bureau (1966). *1960 Censuses of Population and Housing: Procedural History.* Washington, DC: U.S. Government Printing Office.

U.S. Census Bureau (1974). *1970 Censuses of Population and Housing Evaluation and Research Program: Effect of Special Procedures to Improve Coverage in the 1970 Census.* PHC(E)-6. Washington, DC: U.S. Department of Commerce.

U.S. Census Bureau (1976). *Procedural History PHC(R)-1, 1970 Census of Population and Housing.* Washington, DC: U.S. Government Printing Office.

U.S. Census Bureau (1986–1989). *1980 Census of Population and Housing: History.* PHC80-R-2. Issued in parts. Washington, DC: U.S. Census Bureau.

U.S. Census Bureau (1993). *1990 Census of Population and Housing: History—Part A.* 1990 CPH-R-2A. Washington, DC: U.S. Department of Commerce.

U.S. Census Bureau (1995a, October). *1990 Census of Population and Housing: History—Part B.* 1990 CPH-R-2B. Washington, DC: U.S. Department of Commerce.

U.S. Census Bureau (1995b, October). *1990 Census of Population and Housing: History—Part C.* 1990 CPH-R-2C. Washington, DC: U.S. Department of Commerce.

U.S. Census Bureau (1996, March). *1990 Census of Population and Housing: History—Part D.* 1990 CPH-R-2D. Washington, DC: U.S. Department of Commerce.

U.S. Census Bureau (1997). *Results of the 1996 Race and Ethnic Targeted Test.* Population Division Working Paper 18. Washington, DC: U.S. Department of Commerce.

U.S. Census Bureau (1999, January). *Census 2000 Operational Plan Using Traditional Census-Taking Methods.* Washington, DC: U.S. Department of Commerce.

U.S. Census Bureau (2001, June). *Potential Life Cycle Savings for the 2010 Census.* Washington, DC: U.S. Census Bureau.

U.S. Census Bureau (2003, February). *Census 2000 Testing, Experimentation, and Evaluation Program Summary Documentation: Program Modifications Since May 2002.* Washington, DC: U.S. Census Bureau.

U.S. Census Bureau (2005, September). *Estimated Life Cycle Costs for Reengineering the 2010 Decennial Census Program.* Washington, DC: U.S. Census Bureau.

U.S. Census Bureau (2007, February). *U.S. Census Bureau: Budget Estimates, Fiscal Year 2008, Congressional Submission.* Washington, DC: U.S. Census Bureau.

U.S. Census Bureau (2008a, February). *U.S. Census Bureau: Budget Estimates, Fiscal Year 2009, Congressional Submission.* Washington, DC: U.S. Census Bureau.

U.S. Census Bureau (2008b, June). *U.S. Census Bureau: Periodic Censuses and Programs Budget Amendment, Fiscal Year 2009, Congressional Submission.* Washington, DC: U.S. Census Bureau.

U.S. Census Bureau (2009a, April 22). *Letter from Thomas L. Mesenbourg, acting director, to Lawrence D. Brown, Chair, Panel on the Design of the 2010 Census Program of Evaluations and Experiments.* With enclosure: "U.S. Census Bureau Response to The National Academies Committee on National Statistics (CNSTAT) on the 2010 Census Program of Evaluations and Experiments (CPEX) Panel Recommendations." Washington, DC: U.S. Census Bureau.

U.S. Census Bureau (2009b, May). *U.S. Census Bureau: Budget Estimates, Fiscal Year 2010, Congressional Submission.* Washington, DC: U.S. Census Bureau.

U.S. Department of Commerce and Labor (1911). *Reports of the Department of Commerce and Labor, 1910: Report of the Secretary of Commerce and Labor and Reports of Bureaus.* Washington, DC: U.S. Government Printing Office.

U.S. Department of Commerce, Office of Inspector General (2006). *Enumerating Group Quarters Continues to Pose Challenges.* Final Inspection Report No. IPE-18046. Washington, DC: U.S. Department of Commerce.

U.S. General Accounting Office (1994, September). *Decennial Census: 1995 Test Census Presents Opportunities to Evaluate New Census-Taking Methods.* Statement of William M. Hunt before the U.S. House Subcommittee on Census, Statistics, and Postal Personnel. GAO/T-GGD-94-136. Washington, DC: Government Printing Office.

U.S. General Accounting Office (2000, February). *2000 Census: New Data Capture System Progress and Risks.* GAO/AIMD-00-61. Washington, DC: U.S. Government Printing Office.

U.S. Government Accountability Office (2008, July). *2010 Census: Census Bureau's Decision to Continue with Handheld Computers for Address Canvassing Makes Planning and Testing Critical.* Report GAO-08-936. Washington, DC: U.S. Government Printing Office.

U.S. Government Accountability Office (2009a, October 7). *2010 Census: Census Bureau Continues to Make Progress in Mitigating Risks to a Successful Enumeration, but Still Faces Various Challenges.* Report GAO-10-132T. Testimony of Robert Goldenkoff before the Subcommittee on Federal Financial Management, Government Information, Federal Services and International Security, Committee on Homeland Security and Government Affairs, U.S. Senate. Washington, DC: U.S. Government Printing Office.

U.S. Government Accountability Office (2009b, November). *2010 Census: Census Bureau Has Made Progress on Schedule and Operational Control Tools, but Needs to Prioritize Remaining System Requirements.* Report GAO-10-59. Washington, DC: U.S. Government Printing Office.

U.S. Government Accountability Office (2009c, July). *High-Risk Series: Restructuring the U.S. Postal Service to Achieve Sustainable Financial Viability (New).* GAO-09-937SP. Washington, DC: U.S. Government Printing Office.

U.S. Government Accountability Office (2009d, March 5). *Information Technology: Census Bureau Needs to Strengthen Testing of 2010 Decennial Systems.* Report GAO-09-413T. Testimony of David A. Powner and Robert Goldenkoff before the Subcommittee on Federal Financial Management, Government Information, Federal Services and International Security, Committee on Homeland Security and Government Affairs, U.S. Senate. Washington, DC: U.S. Government Printing Office.

Weinberg, D. H. (2008). Early thoughts on a research and testing strategy for the 2020 census. Presentation to the Panel on the Design of the 2010 Census Program of Experiments and Evaluations, November 10, 2008, Washington, DC.

Wells, R. V. (2000). 1790 census. In M. J. Anderson (Ed.), *Encyclopedia of the U.S. Census*, pp. 115–116. Washington, DC: CQ Press.

West, K. K., J. G. Robinson, and M. Bentley (2005). Did proxy respondents cause age heaping in the census 2000? In *Proceedings of the Section on Survey Research Methods*, pp. 3658–3665. Alexandria, VA: American Statistical Association.

Whitford, D. C. (1996). The 1996 Integrated Coverage Measurement test. In *Proceedings of the Section on Survey Research Methods*, pp. 389–393. Alexandria, VA: American Statistical Association.

Whitworth, E. (2002, August 14). *Internet Data Collection*. Census 2000 Evaluation A.2.b. Washington, DC: U.S. Census Bureau.

Williams, P. (2006). *The Australian 2006 census and the Internet.* Working Paper 9, Work Session on Electronic Raw Data Reporting. Geneva: United Nations Statistical Commission and Economic Commission for Europe and Conference of European Statisticians.

# Biographical Sketches of Panel Members and Staff

**Lawrence D. Brown** *(Chair)* is Miers Bush professor in the Department of Statistics of the Wharton School at the University of Pennsylvania. He is a fellow of the American Statistical Association, a fellow and past president of the Institute of Mathematical Statistics, and a member of the National Academy of Sciences. At the National Research Council (NRC), he has served on the Committee on National Statistics (CNSTAT) and its Panel to Review the 2000 Census and Panel on Correlation Bias and Coverage Measurement in the 2010 Decennial Census. He also served on the NRC's Commission on Physical Sciences, Mathematics, and Applications and its Board on Mathematical Sciences. He was a critic of the Census Bureau's plans to incorporate sampling in the census. He has a B.S. from the California Institute of Technology and a Ph.D. from Cornell University.

**Richard A. Berk** is professor of criminology and statistics at the University of Pennsylvania. He is active regarding a range of methodological concerns, such as causal inference, statistical learning, and methods for evaluating social programs. His main areas of research include the inmate classification and placement systems, law enforcement strategies for reducing domestic violence, the role of race in capital punishment, detecting violations of environmental regulations, claims that the death penalty serves as a general deterrent, and forecasting short-term changes in urban crime patterns. Currently, he is working on the development and application of statistical learning procedures for data sets in the behavioral, social, and economic sciences. He has previously served on the faculties of Northwestern University and the University of California, Santa Barbara, and joined the University of Pennsylvania faculty after serving as professor, director of the Center for the Study of the Environment and Society, and director of the Statistical Consulting Center at the University of California, Los Angeles. He has been elected to the Sociological Research Association and is a fellow of the American Association for the Advancement of Science, the American Statistical Association, and the Academy of Experimental Criminology. He was awarded the Paul S. Lazarsfeld Award for methodological contributions from the American Sociological Association. At the NRC, he has served on the Committee on Applied and Theoretical Statistics, the Panel on Monitoring the Social Impact of the AIDS Epidemic, the Working Group on Field Experimentation in Criminal Justice, and the Panel on Sentencing. He has a B.A. from Yale University and a Ph.D. from Johns Hopkins University.

**Eric T. Bradlow** is K.P. Chao professor of marketing, statistics, and education at the Wharton School of the University of Pennsylvania. He previously held positions at the Educational Testing Service and at E.I. DuPont de Nemours and Company. He has won numerous teaching awards, and his research interests include Bayesian modeling, statistical computing, and developing new methodology for unique data structures. His current projects center on optimal resource allocation, choice modeling, and complex latent structures. He serves as associate editor for the *Journal of Computational and Graphical Statistics, Marketing Science, Quantitative Marketing and Economics*, and *Psychometrika*, and as senior associate editor for the *Journal of Educational and Behavioral Statistics*. He is a fellow of the American Statistical Association. At the NRC, he served on CNSTAT's Panel to Review the U.S. Department of Agriculture's Measurement of Food Insecurity and Hunger. He has a B.S. from the University of Pennsylvania, an A.M. from Harvard University, and a Ph.D. in mathematical statistics from Harvard University.

**Michael L. Cohen** *(co-study director)* is a senior program officer for the Committee on National Statistics. He has served as study director or program officer for numerous CNSTAT census panels, as well as a series of workshops on statistical topics and applications in defense testing and acquisition. Formerly, he was a mathematical statistician at the Energy Information Administration, an assistant professor in the School of Public Affairs at the University of Maryland, and a visiting lecturer in statistics at Princeton University. His general area of research is in the use of statistics in public policy, with particular interest in census undercount, model validation, and robust estimation. He is a fellow of the American Statistical Association. He has a B.S. in mathematics from the University of Michigan and M.S. and Ph.D. degrees in statistics from Stanford University.

**Daniel L. Cork** *(co-study director)* is a senior program officer for the Committee on National Statistics, currently serving as study director of the Panel to Review the 2010 Census. He joined the CNSTAT staff in 2000 and has served as study director or program officer for several census panels, including the Panels on Residence Rules in the Decennial Census, Research on Future Census Methods (2010 Planning panel), and Review of the 2000 Census. He also directed the Panel to Review the Programs of the Bureau of Justice Statistics (in cooperation with the Committee on Law and Justice) and was senior program officer for the Panel on the Feasibility, Accuracy, and Technical Capability of a National Ballistics Database (joint with the Committee on Law and Justice and the National Materials Advisory Board). His research interests include quantitative criminology, geographical analysis, Bayesian statistics, and statistics in sports. He has a B.S. in statistics from George Washington University and an M.S. in statistics and a joint Ph.D. in statistics and public policy from Carnegie Mellon University.

**Ivan P. Fellegi** is chief statistician emeritus of Canada, having served as chief statistician from 1985 to 2008. He joined Statistics Canada (then the Dominion Bureau of Statistics) in 1957, serving as director of sampling research and consultation and director general of methodology and systems, assistant chief statistician, and deputy chief statistician before his appointment as chief statistician. He has published extensively in the areas of census and survey methodology, in particular on consistent editing rules and record linkage. A past chair of the Conference of European Statisticians of the United Nations Economic Commission for Europe, he is an honorary member and past president of the International Statistical Institute, an honorary fellow of the Royal Statistical Society, past president of the International Association of Survey Statisticians, and past president and Gold Medal recipient of the Statistical Society of Canada. He was made Member of the Order of Canada in 1992 and promoted to Officer in 1998 and has received the nation's Outstanding Achievement Award; he has also provided advice on statistical matters to his native Hungary following its transition to democracy and, in 2004, was awarded the Order of Merit of the Republic of Hungary. At the National Research Council, he was a member of the Panel on Privacy and Confidentiality as Factors in Survey Response, the Panel on Census Requirements in the Year 2000 and Beyond, the Panel on Decennial Census Methodology, and the Panel on the Design of the 2010 Census Program of Experiments and Evaluations. He has a B.Sc. from the University of Budapest and M.Sc. and Ph.D. degrees in survey methodology from Carleton University.

**Linda Gage** is senior demographer in the State of California's Demographic Research Unit. Her primary objective is to improve the currency, completeness, and accuracy of official state and federal demographic data that portray the people of California. She is actively involved in producing and evaluating intercensal population estimtes for California and assessing data from the American Community Survey. She represents the state's demographic program and interests in federal and professional forums and evaluates the effect of various demographic and statistical programs on the state. She chairs the steering committee of the Census Bureau's Federal-State Cooperative Program for Population Estimates, serves on the Population Association of America (PAA) Public Affairs Committee and Committee on Population Statistics, and represents PAA on the Census Advisory Committee of Professional Associations. She served as the Governor's Liaison for Census 2000 and represented the State Data Center network and the Population Association of America on the U.S. Secretary of Commerce's Decennial Census Advisory Committee. She has B.A. and M.A. degrees in sociology, with emphasis in demography, from the University of California, Davis.

**Vijay Nair** is Donald A. Darling professor of statistics and professor of industrial and operations engineering at the University of Michigan. He has been chair of the statistics department since 1998. He was a research scientist at Bell Laboratories for 15 years before joining the faculty at Michigan. His area of expertise is engineering statistics, including quality and productivity improvement, experimental design, reliability, and process control. He is a fellow of the American Association for the Advancement of Science, the American Statistical Association, and the Institute of Mathematical Statistics, and an elected member of the International Statistical Institute. He is a former editor of *Technometrics* and *International Statistical Review* and has served on many other editorial boards. He is currently the chair of the Board of Trustees of the National Institute of Statistical Sciences and is a member of CNSTAT. At the NRC, he has served on several panels, including the Panel on Statistical Methods for Testing and Evaluating Defense Systems and the Assessment Panel on NIST's Information Technology Center, and chaired the Oversight Committee for the Workshop on Testing for Dynamic Acquisition of Defense Systems. He has a Ph.D. in statistics from the University of California, Berkeley.

**Jesse H. Poore, Jr.,** holds the Ericsson/Harlan D. Mills chair in software engineering in the Department of Electrical Engineering and Computer Science at the University of Tennessee, Knoxville. He is also director of the University of Tennessee–Oak Ridge National Laboratory Science Alliance, a program to promote and stimulate joint research between those two organizations. He conducts research in cleanroom software engineering and teaches software engineering courses. He has held academic appointments at Florida State University and the Georgia Institute of Technology, served as a National Science Foundation rotator, worked in the Executive Office of the President, and was executive director of the Committee on Science and Technology in the U.S. House of Representatives. He is a member of the Association for Computing Machinery and the Institute of Electrical and Electronics Engineers, and a fellow of the American Association for the Advancement of Science. At the NRC, he served on the CNSTAT Panel on Statistical Methods for Testing and Evaluating Defense Systems and the oversight committee for the Workshop on Testing for Dynamic Acquisition of Defense Systems. He has a Ph.D. in information and computer science from the Georgia Institute of Technology.

**Nora Cate Schaeffer** is professor of sociology at the University of Wisconsin–Madison. Her areas of expertise include respondent behavior and interviewer-respondent interaction. Her past research has concentrated on a number of different areas in survey methodology dealing with non-sampling error, both nonresponse and response errors of various kinds. She was on the editorial board of *Public Opinion Quarterly*, *Sociological Methodology*, and *Sociological Methods Research*. A past member of the Committee on National Statistics, she also served on the Panel to Evaluate Alternative Census Methods. She has an A.B. from Washington University and a Ph.D. in sociology from the University of Chicago.

**Allen L. Schirm** is a vice president and director of human services research at Mathematica Policy Research. Formerly, he was Andrew W. Mellon assistant research scientist and assistant professor at the University of Michigan. In addition to census methods, his principal research interests include small-area estimation and evaluation design, with application to studies of child well-being and welfare, food and nutrition, and education policy. At the NRC, he has served as a member of the Panel on Research on Future Census Methods, the Panel on Formula Allocations, and the Panel on Estimates of Poverty for Small Geographic Areas, and is currently chairing the Panel on Estimating Children Eligible for School Nutrition Programs Using the American Community Survey. He is a fellow of the American Statistical Association. He received an A.B. in statistics from Princeton University and a Ph.D. in economics from the University of Pennsylvania.

**Judith A. Seltzer** is professor of sociology at the University of California, Los Angeles. Previously, she was on the faculty of the University of Wisconsin–Madison, where she contributed to the development and implementation of the National Survey of Families and Households. Her research interests include kinship patterns, intergenerational obligations, relationships between nonresident fathers and children, and how legal institutions and other policies affect family change. She was part of a cross-university consortium to develop new models for explaining family change and variation and a member of the design team for the Los Angeles Family and Neighborhood Survey. At the NRC, she has served on CNSTAT's Panel on Residence Rules in the Decennial Census and is a member of the Panel to Review the 2010 Census. She has master's and Ph.D. degrees in sociology from the University of Michigan.

**Stanley K. Smith** is professor of economics and director of the Bureau of Economic and Business Research (BEBR) at the University of Florida. He is also director of BEBR's population program, which produces the official state and local population estimates and projections for the state of Florida. He is Florida's representative to the Federal-State Cooperative Program for Population Estimates and Projections and a past president of the Southern Demographic Association. He has also served on the U.S. Census Bureau's Decennial Advisory Committee. His research interests include the methodology and analysis of population estimates; he has done particular work on the measurement of seasonal populations. At the NRC, he served on CNSTAT's Panel on Alternative Census Methodologies. He has a degree in history from Goshen College and a Ph.D. in economics from the University of Michigan.

**John H. Thompson** is president of the National Opinion Research Center (NORC) at the University of Chicago. Prior to his appointment as president, he was executive vice president for survey operations, in which capacity he provided oversight and direction for NORC's Economics, Labor Force, and Demography Research Department and the Statistics and Methodology Department. He also served as project director for the National Immunization Survey, conducted on behalf of the Centers for Disease Control and Prevention from November 2004 through July 2006. He joined NORC following a 27-year career at the U.S. Census Bureau, culminating in service as principal associate director for programs. As associate director for decennial census (1997–2001) and chief of the Decennial Management Division (1995–1997), he was the chief operating officer of the 2000 census, overseeing all aspects of census operations. In this capacity, he also chaired the Bureau's Executive Steering Committee for Accuracy and Coverage Evaluation Policy, an internal working group tasked to provide guidance to the director of the Census Bureau and the secretary of commerce concerning statistical adjustment of 2000 census figures. He has received a Presidential Rank Award of Meritorious Executive and Gold, Silver, and Bronze Medals from the U.S. Department of Commerce. He is a fellow of the American Statistical Association. He has bachelor's and master's degrees in mathematics from Virginia Polytechnic Institute and State University.

**Roger Tourangeau** is director of the Joint Program in Survey Methodology at the University of Maryland and a senior research scientist at the University of Michigan. Previously, he was a senior methodologist at the Gallup Organization, where he designed and selected samples and carried out methodological studies; he also founded and directed the Statistics and Methodology Center of the National Opinion Research Center. His research focuses on attitude and opinion measurement and on differences across methods of data collection; he also has extensive experience as an applied sampler and has conducted work on the cognitive aspects of survey methodology. A fellow of the American Statistical Association, he has served on the editorial board of *Public Opinion Quarterly* and on Census Bureau advisory panels. At the NRC, he is currently a member of CNSTAT and previously served on the Panel on Residence Rules in the Decennial Census. He has a Ph.D. in psychology from Yale University.

**Kirk Wolter** is senior fellow and director of the Center for Excellence in Survey Research at the National Opinion Research Center, where he has also served as senior vice president for statistics and methodology. He is also professor of statistics, part time, at the University of Chicago. During his career, he has led or participated in the design of many of America's largest information systems, including the Current Business Surveys, the Current Employment Statistics program, the Current Population Survey, the 1980 and 1990 decennial censuses, the National 1997 Longitudinal Survey of Youth, and the National Resources Inventory. He is a fellow of the American Statistical Association, an elected member of the International Statistical Institute, and past president of the International Association of Survey Statisticians and of the Survey Research Methods section of the American Statistical Association. At the NRC, he served on CNSTAT's Panel on Conceptual, Measurement, and Other Statistical Issues in Developing Cost of Living Indexes and the Panel on Measuring Business Formation, Dynamics, and Performance. He has an M.A. and a Ph.D. in statistics, both from Iowa State University.

## COMMITTEE ON NATIONAL STATISTICS